ENERGY SCIENCE, ENGINEERING AND TECHNOLOGY

NEW NUKES

ASSESSMENTS OF LIGHT WATER SMALL MODULAR AND ADVANCED REACTOR CONCEPTS

ENERGY SCIENCE, ENGINEERING AND TECHNOLOGY

ENERGY SCIENCE, ENGINEERING AND TECHNOLOGY

NEW NUKES

ASSESSMENTS OF LIGHT WATER SMALL MODULAR AND ADVANCED REACTOR CONCEPTS

DARNELL WALTERS
EDITOR

nova publishers
New York

Library of Congress Cataloging-in-Publication Data

ISBN: 978-1-63484-553-3

Published by Nova Science Publishers, Inc. † New York

CONTENTS

PREFACE

Energy demand in the United States is expected to continue to grow over the coming decades, and the Department of Energy (DOE) considers nuclear energy to be one way to help meet this increased demand without producing air pollution. However, the current domestic commercial nuclear reactor fleet, consisting of 99 large light water reactors (LWRs), that provide about 20 percent of U.S. electricity, is aging, and some reactors have shut down in recent years. LWRs use light, or ordinary, water to cool the reactor. New reactor concepts are under development as alternative energy options. Light water small modular reactors (SMRs) have some similarities, including the coolant used, to the existing large LWRs, and advanced reactors differ more from the large LWRs. Both new reactor concepts differ from the existing large LWRs in potential applications. This book discusses the status of light water SMR and advanced reactor concepts under development; the intended benefits of these new reactor concepts; and the challenges associated with developing and deploying these new types of reactors. Furthermore, the book describes DOE's Office of Nuclear Energy's (NE) approach to advanced nuclear reactor research and development (R&D) and examines how NE plans and prioritizes its advanced reactor R&D activities, including deploying an advanced reactor.

In: New Nukes
Editor: Darnell Walters

ISBN: 978-1-63484-553-3

Chapter 1

NUCLEAR REACTORS: STATUS AND CHALLENGES IN DEVELOPMENT AND DEPLOYMENT OF NEW COMMERCIAL CONCEPTS*

United States Government Accountability Office

WHY GAO DID THIS STUDY

Energy demand in the United States is expected to continue to grow over the coming decades, and Department of Energy (DOE) considers nuclear energy to be one way to help meet this increased demand without producing air pollution. However, the current domestic commercial nuclear reactor fleet, consisting of 99 large light water reactors (LWRs) that provide about 20 percent of U.S. electricity, is aging, and some reactors have shut down in recent years. LWRs use light, or ordinary, water to cool the reactor. New reactor concepts are under development as alternative energy options. Light water small modular reactors (SMRs) have some similarities, including the coolant used, to the existing large LWRs, and advanced reactors differ more from the large LWRs. Both new reactor concepts differ from the existing large LWRs in potential applications.

* This is an edited, reformatted and augmented version of a United States Government Accountability Office publication, No. GAO-15-652, dated July 2015.

GAO was asked to conduct a technology assessment of these new reactor concepts in the United States. This report discusses (1) the status of light water SMR and advanced reactor concepts under development; (2) the intended benefits of these new reactor concepts; and (3) the challenges associated with developing and deploying these new types of reactors. GAO reviewed documents from DOE and NRC, and interviewed DOE and NRC staff as well as industry representatives involved in developing reactors. GAO, with the assistance of the National Academies, convened a meeting with a group of 20 experts on nuclear reactor development and related issues to provide additional information.

WHAT GAO FOUND

In the United States, four light water small modular reactors (SMRs)—nuclear power reactors with a generating capacity of less than 300 MW of electricity—have been developed to the point that the reactor designers have begun discussing design certification and license applications with the Nuclear Regulatory Commission (NRC), and one SMR designer has established time frames for applications to NRC and construction of a power plant. The Department of Energy (DOE) has provided financial support to the designers of two SMRs for reactor certification and licensing work. DOE supports the SMR design by NuScale through a cost-sharing agreement in which DOE will pay as much as half of NuScale's costs—up to $217 million over 5 years—for certifying the design. The SMR design by mPower has a similar cost-sharing agreement with DOE, but DOE is no longer providing funds because mPower has scaled back its efforts while it looks for additional investors. NuScale expects to submit a design certification application to NRC in late 2016, with its first power plant beginning operation as early as 2023. Other SMR designers do not yet have established time frames for such applications. DOE also supports research and development (R&D) activities on advanced reactor concepts that focus on the high temperature gas reactor and the sodium fast reactor. DOE provides this support in areas such as fuels and material qualification and reactor safety studies. DOE and NRC officials do not expect applications for advanced reactors for at least 5 years.

According to DOE officials and reactor designers, both SMRs and advanced reactors are intended to provide benefits that could facilitate the use of nuclear reactors in new markets or commercial applications. SMR designers plan to decrease the overall cost and time for reactor construction, compared with existing large light water reactors (LWRs), without significantly increasing ongoing operational costs. They told GAO they expect that the smaller size of SMRs may expand the locations where a nuclear power plant could be constructed. For example, they may be used in remote or rural areas that have lower electricity demands or smaller distribution systems. DOE officials and reactor designers expect advanced reactors to operate at higher temperatures and therefore they could generate electricity more efficiently. Furthermore, they told GAO heat from these higher temperature reactors could be used directly in certain industrial processes that currently depend on fossil fuels. Some advanced reactors may also allow for improved spent nuclear fuel recycling and management.

DOE officials and SMR and advanced reactor designers told GAO they face challenges in developing and deploying these reactors. SMR designers face technical challenges in demonstrating economic feasibility and safety without increasing reactor complexity, and advanced reactor designers face greater technical challenges because advanced reactors differ more from current reactors than SMRs. Reactor designers told GAO they face challenges associated with the up to $1 billion to $2 billion cost of developing and certifying a design. Even with a reactor design ready to submit to NRC, the licensing and construction can take nearly a decade or more before a reactor is operational. DOE officials, members of GAO's expert group, and reactor designers said that the cost and time needed to certify or license a reactor design and construct it, along with uncertainty about the energy market in the future and potential customer interest, create obstacles to the development and deployment of new reactors.

ABBREVIATIONS

AEO2015	Annual Energy Outlook 2015
CFR	Code of Federal Regulations
COL	Combined License
DC	Design Certification
DOE	Department of Energy
EIA	Energy Information Administration
FHR	Fluoride Salt Cooled High Temperature Reactor
FOAK	First-of-a-Kind
GIF	Generation IV International Forum
HTGR	High Temperature Gas Cooled Reactor

MSR	Molten Salt Reactor
LWR	Light Water Reactor
MWe	Megawatt-Electric
MWth	Megawatt-Thermal
NGCC	Natural Gas-Fired Combined Cycle
NGNP	Next Generation Nuclear Plant
NOAK	Nth-of-a-Kind
NRC	Nuclear Regulatory Commission
O&M	Operations and Maintenance
R&D	Research and Development
SFR	Sodium Cooled Fast Reactor
SMR	Small Modular Reactor
TVA	Tennessee Valley Authority

* * *

July 28, 2015

The Honorable Senator Dianne Feinstein
Ranking Member
Subcommittee on Energy and Water Development
Committee on Appropriations
United States Senate

Dear Senator Feinstein:

Over the coming decades, energy demand in the United States is expected to continue to grow, and energy security, scalability, reliability, and the mitigation of greenhouse gas emissions from the burning of fossil fuels, such as coal, will remain key elements of national interest. The Department of Energy (DOE) considers nuclear energy to be a proven technology that can reliably generate large amounts of electricity without producing air pollution or greenhouse gases. The U.S. agency that regulates commercial nuclear reactors, the Nuclear Regulatory Commission (NRC), has been granting license renewals for the continued operation of the current commercial fleet of large light water nuclear reactors to extend their operating lifetimes to 60 years, and may grant another license renewal to allow operation for 80 years if a reactor operator applies for one and NRC determines through a review of the application that the additional license renewal is acceptable. However, the industry faces several challenges. Commercial nuclear reactors in the United States are aging, and some have shut down before their license expirations because of economic pressures in certain markets. In addition, the accident at Japan's Fukushima Daiichi commercial nuclear power plant, which was damaged by the March 2011 earthquake and tsunami, has increased concerns about nuclear safety and resulted in modifications to some reactors that have affected their economic competitiveness. There are also concerns and uncertainties surrounding long-term disposal options for spent nuclear fuel as well as concerns over nuclear proliferation and terrorism.[1]

Government agencies and industry sponsor and conduct research on new nuclear reactor concepts, which are intended to provide additional capabilities and improvements, such as improved efficiency or simpler reactor design, over the existing fleet of large commercial light water nuclear reactors (LWRs).[2] In the United States, these large LWRs have an average generating capacity over 1,000 megawatt-electric (MWe).[3] From funds appropriated to DOE for nuclear energy in fiscal year 2015, DOE used $152.5 million to support small modular reactors (SMRs) and advanced reactor concepts. Of that amount, DOE allocated $54.5 million to support the industry's reactor designers with their licensing work on light water SMRs, which are small reactors under 300 MWe that are designed for modular production and assembly with components transportable by road, rail, or barge.[4] DOE's support of light water SMRs, which are a type of LWR that are less than about a third the size of average-sized large LWRs, is in the form of industry cost-sharing agreements, by which DOE provides funds to match those expended by its industry partners to support the certification and licensing of specific light water SMR designs. According to DOE documents and reactor designers, the goal for SMR designs is to provide a more commercially flexible reactor option with lower investment cost. DOE also allocated $98 million to support research and development (R&D) on advanced reactors, which use coolant technologies significantly different from LWRs. As we previously found, DOE supports research on advanced reactor concepts with the goal of improving the economic competitiveness of nuclear technology relative to other energy options, ensuring that nuclear energy continues to play a role in meeting our nation's energy needs, minimizing the risks of nuclear proliferation, and addressing environmental challenges such as greenhouse gas emissions.[5]

You asked us to conduct a technology assessment of new reactor concepts under development in the United States. This report discusses (1) the status of light water SMR and advanced reactor concepts under development; (2) the intended benefits of these new reactor concepts; and (3) the challenges associated with developing and deploying these new types of reactors.

To address these objectives, we reviewed relevant NRC regulations as well as our past work. We reviewed reports and technical literature from DOE, international nuclear technology organizations, and reactor designers that describe DOE's and industry's efforts to develop light water SMRs and advanced reactor concepts, and we attended six conferences on SMR or advanced reactor concepts. In addition, to understand the status of light water SMRs and advanced reactor concepts under development, their applications, and challenges they face, during this and related work we reviewed reports and economic studies and interviewed DOE officials, NRC officials, and industry and reactor operator representatives, including designers of light water SMRs and advanced reactors that receive DOE support or have been in pre-licensing discussions with NRC. We also visited a DOE national laboratory—Idaho National Laboratory—to better understand the R&D activities DOE has undertaken with regard to light water SMRs and advanced reactor concepts. With the assistance of the National Academies, we convened a meeting of 20 experts on nuclear reactor technology and related issues to obtain additional information and advice for this review. These experts were selected from academia, government, and industry, with expertise ranging from reactor and electricity economics to reactor development and licensing. We have limited the scope of our review to specific domestic light water SMR designs and the advanced reactor concepts that DOE has focused on for domestic R&D because these are the new reactor technologies closest to being certified, licensed, and built in the United States.

We conducted our work from June 2014 to July 2015 in accordance with all sections of GAO's quality assurance framework that are relevant to technology assessments. The framework requires that we plan and perform the engagement to obtain sufficient and appropriate evidence to meet our stated objectives and to discuss any limitations to our work. We believe that the information and data obtained, and the analysis conducted, provide a reasonable basis for our findings and conclusions in this product.

1. Background

Nuclear reactors, along with other types of power plants, provide electricity to consumers in the United States, but nuclear power plants can face economic pressures that lead to their shutting down before the expiration of their operating licenses, and other large power plants can face similar economic pressures. Designing and certifying a new type of nuclear reactor design can cost up to $1 billion to $2 billion, with much of the cost going to R&D and reactor design work, and around $50 million to $75 million paying for NRC's fees for design certification.[6] The design work and design certification is dependent on funding and can take several years, including up to 10 years or more for design work before submitting an application to NRC and nearly 3.5 years, as a best-case scenario, for certifying a LWR design.[7] With a certified design in place, the decision by a customer[8] to build a new nuclear reactor of any type is ultimately an economic one, requiring consideration of a number of factors, including reactor license approval times of at least 4 to 6 years, reactor construction costs, large LWR construction times of 6 years or more, and ongoing operation and maintenance costs. While some of these time frames can have overlapping schedules, it is still a multi-decade process to design, license, and build a reactor using a new design.

1.1. Nuclear Reactors Are One Component of the Larger Electric Grid

Nuclear power plants and other electricity production facilities generate electricity using fuels such as uranium, coal, natural gas, and renewable energy sources. This electricity is sent through the electric grid, which consists of high-voltage, high-capacity transmission systems, to areas where it is transformed to a lower voltage and sent through the local distribution system for use by business and residential consumers. During this process, a grid operator must constantly balance the generation and consumption of electricity. To do so, grid operators monitor electricity consumption from a centralized location using computerized systems and send minute-by-minute signals to power plants to adjust their output—to the extent possible for each type of plant—to match changes in the demand for electricity. Electricity demand can vary throughout a day, as well as seasonally, so grid operators use baseload plants and peaker plants.[9] Historically, baseload plants, typically nuclear or coal powered, generally cost more to build but supply electricity at a lower hourly cost. Peaker plants, such as natural gas facilities, can be built relatively cheaply and generally operate at a higher hourly cost, and they can rapidly be brought on or offline in response to changes in electricity demand.[10] Some renewable energy sources, such as solar and wind power, are intermittent and are prioritized as electricity sources by grid operators because of green energy usage requirements and their near-zero operating costs when that energy source is available. The intermittent nature of electricity generated by these renewable energy sources can be a challenge for baseload plant operators—including nuclear power plants operators—because the operators cannot increase and decrease their power production to balance it with that of intermittent electricity sources without additional wear on their equipment or less economic power generation. Another difference between intermittent renewables, like wind and solar, and nuclear power or fossil fuel plants is that they are not generally dispatchable without battery storage, meaning that they are either producing power or not without respect to demand.

In the last several years, five commercial nuclear reactors have shut down before their operating licenses expired, and additional reactors currently face economic pressures that may lead to early shutdowns. These economic pressures vary by market, but include competition from renewable energy sources, or less expensive natural gas, coupled with increased reactor operating costs.[11] As of April 2015, the commercial nuclear reactor fleet in the United States consisted of 99 operating nuclear reactors, which provided nearly 20 percent of total U.S. electricity generation. Even if these reactors operate for their full license durations, assuming all the reactors receive license extensions allowing operational lifetimes of 60 years, those licenses will begin expiring in 2029 unless reactor operators apply for, and NRC determines through a review of the applications to allow an additional 20-year renewal period to extend allowed reactor operational lifetimes to 80 years. Additional reactor shutdowns would require increased electricity generation from other sources, including fossil fuels and renewable energy sources, or would require improvements in energy efficiency. As shown in figure 1, U.S. electricity demand over the next several decades is forecast to increase, so additional sources of electricity generation, such as nuclear power, would be needed to meet that demand, according to the U.S. Energy Information Administration (EIA) Annual Energy Outlook 2015 (AEO2015) Reference Case.[12]

1.2. Designing and certifying or licensing a new reactor

In the United States, both DOE and industry play major roles in R&D for new reactor concepts, and NRC certifies the resulting designs and licenses the construction and operation of the reactors built to those designs. The overall process of developing and certifying a specific reactor design can take 10 years or more for design work and nearly 3.5 years, as a best case, for certification, and includes the following elements:

- the R&D by reactor designers to support their claims about reactor safety and economic competitiveness, including construction costs and ongoing operations and maintenance costs;
- engineering design work by reactor designers including any needed fuel and materials development and qualification; and
- reactor design certification efforts by the designers and the NRC.

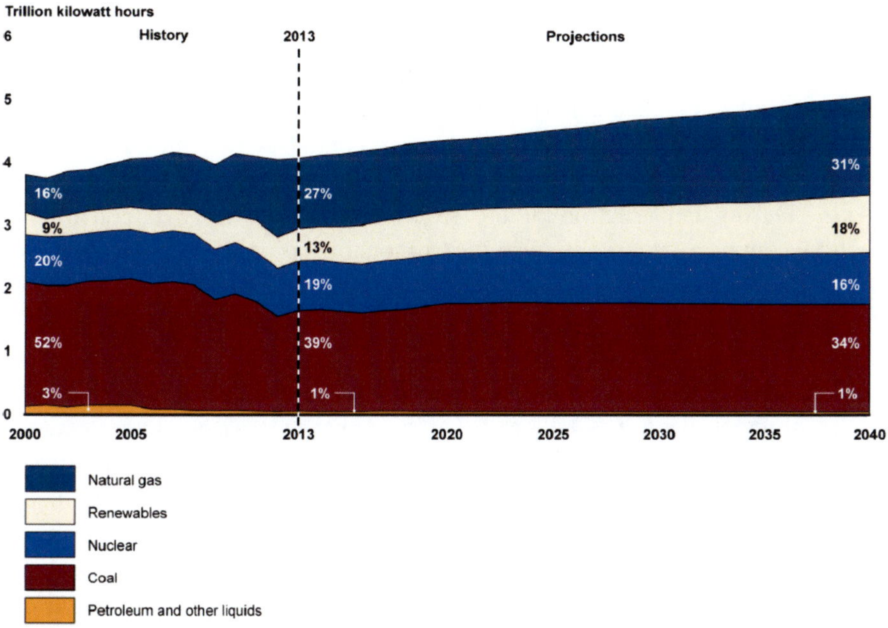

Source: GAO, adapted from U.S. Energy Information Administration Annual Energy Outlook 2015 Reference Case. | GAO-15-652.

Note: This energy forecast, by fuel type, incorporates a number of assumptions, including the effects of planned and unplanned builds and retirements of nuclear reactors. It also assumes that most existing reactor operators will apply for and receive approval from the Nuclear Regulatory Commission for a second 20-year license renewal.

Figure 1. U.S. Energy Information Administration Annual Energy Outlook 2015 Reference Case historical data and forecast of electricity generation by fuel type (trillion kilowatt hours).

While the R&D and engineering work can vary based on the specific design, the NRC certification and licensing process is defined through federal regulations. However, according to NRC documents, this process is focused on LWR designs and may require adjustments, such as exemptions from the current process, when applied to advanced reactors.

NRC has two licensing paths for constructing and operating nuclear power plants defined in the Code of Federal Regulations (CFR). The first follows 10 CFR Part 50 (Part 50). Under Part 50, a license for a proposed power plant is issued in two parts—a construction permit allowing the plant construction to begin, and an operating license. This licensing process requires several items in the application to be reviewed by NRC and its independent advisory committee before a reactor is built and operational. These items include the plant design and operations, environmental and site studies, and emergency response plans. Once the review is complete, the NRC prepares a Safety Evaluation Report, which summarizes the potential effects on public health, safety, and the environment based on the final design and location of the reactor. During construction of the reactor, NRC verifies that construction meets acceptance criteria to provide reasonable assurance the reactor will operate in conformance with the license and regulations. In addition, NRC provides opportunities for public hearings or comments at specified points.

The second path, following 10 CFR Part 52 (Part 52), requires the same types of studies and reports as the Part 50 process, but Part 52 allows for a single combined license (COL) to be granted that authorizes the construction and operation of a specific reactor design at a specific site, subject to post-construction verification that the plant meets the acceptance criteria defined in the COL. Light water reactor designers we interviewed may use Part 52 because it also allows for two options that can be exercised independent of a construction permit or license: (1) an early site permit for NRC approval of a site for a future reactor, and (2) a design certification (DC) for NRC approval of a standardized nuclear reactor design. According to NRC documents, a Part 52 COL application can reference an approved early site permit or DC to increase regulatory efficiency and predictability in the review process. Using this process, reactor designers may apply for a DC for their specific design so that it may be incorporated into the COL applications that potential customers would submit, reducing the timeframes for the COL application review. This use of a single DC for multiple reactor COL applications is in contrast with the Part 50 process, where the reactor design would need to be submitted and approved by NRC for each proposed construction site. However, members of our expert group noted that the Part 50 process may be more advantageous for a reactor customer who is constructing and licensing a new reactor for the first time, because fewer design details need to be complete when construction begins and design changes can be easier to make than amending designs certified through Part 52. For these reasons, advanced reactor designers and their customers may be particularly interested in Part 50, according to members of our expert group. Regardless of whether a reactor designer or its customer pursues Part 50 or Part 52, the NRC regulatory process for a reactor can take at least 4-6 years before a reactor is licensed to be built and operated. NRC estimates the DC process for an LWR design will take nearly 3.5 years as a best-case scenario, assuming the submission of a high-quality application and that NRC's requests for additional information from the reactor designers are received and responded to in a timely manner and meet NRC needs. Recent LWR design certifications—for the Westinghouse AP1000 and the GE Hitachi ESBWR—have taken about 15 and 11 years, including revisions.[13]

NRC encourages reactor designers and license applicants to engage in pre-application discussions with NRC to help identify potential certification or licensing issues the designers may wish to address before submitting DC or COL applications, according to NRC documents and officials and reactor designers. NRC officials told us identifying such issues is important for new light water SMR and advanced reactor concepts because these concepts incorporate design features that differ from those of existing large LWRs—for example, the new reactor concepts may approach reactor safety differently from the large LWRs that NRC has experience in licensing. If a designer intends to submit a DC application, these pre-application discussions may also lead to NRC drafting a Design Specific Review Standard, an NRC document that can be used to guide NRC's review of the application. According to NRC documents, before submitting DC applications, reactor designers may also request NRC to conduct pre-application readiness reviews to provide general feedback on the readiness of the applications for submission. During the review process, reactor designers told us they must expend resources to conduct studies and provide responses to NRC inquiries to support DC application claims about the reactor design. According to NRC regulations, reactor designers must also reimburse NRC for the agency's staff time during pre-application discussions, pre-application readiness reviews, and application reviews.[14]

In this report, most references to the licensing process describe the Part 52 process because the reactor designers with the earliest planned license approval activities told us they intend to submit DC applications. As noted above, the first step in this process may consist of reactor designers working with NRC in optional pre-application discussions. According to NRC documents, as the designers get closer to submitting a DC application to NRC, a potential customer for the reactor may begin conducting early site characterization, with designer assistance, to prepare for a COL application for a specific reactor design and site. Alternatively, a potential reactor customer may plan to submit an early site permit application independent of a specific reactor design, which would support a later COL application for any reactor design meeting the parameters of the permit at that site. According to NRC documents, a DC application and COL application may then be processed in parallel by the NRC. If the applications are successful, the end result would be a reactor design that is certified by the NRC independent of any specific site along with a COL that permits that reactor design to be built and operated at a specific site. See figure 2 for the notional best-case time frames associated with this process which, according to NRC documents, assume well-prepared applications, timely review, and timely responses to NRC requests for additional information. The time frames shown in figure 2 are based on NRC best-case estimates for a light water reactor design. Because of the need for more adjustments and exemptions to apply these processes to an advanced reactor design, time frames for an advanced reactor design would be longer, according to DOE and NRC officials and members of our expert group.

1.3. Reactor Economics

There are several considerations for a customer, such as a utility, planning to construct and operate a nuclear power plant. Besides considering atmospheric emissions, electricity reliability, and source diversification (that is, having a portfolio of electricity production methods), a customer will also consider whether the power plant will be cost-competitive with other sources of electricity.[15]

For example, one type of customer, traditionally called a utility, has a guaranteed rate of return set by a public authority to meet a baseload demand so that the utility can fully recover its prudently incurred costs, including the cost of its debt, plus an established return on its equity. These customers may have an easier path to nuclear power plant construction because, once the plant is approved by the public authority, the customer may be guaranteed a certain return on its investment. On the other hand, a customer that is not guaranteed a rate by a public authority—a merchant generator—will have to pay for the plant by competing with other electricity generators or through prior purchase agreements, which are guaranteed rates over a contracted time from an electricity purchaser.

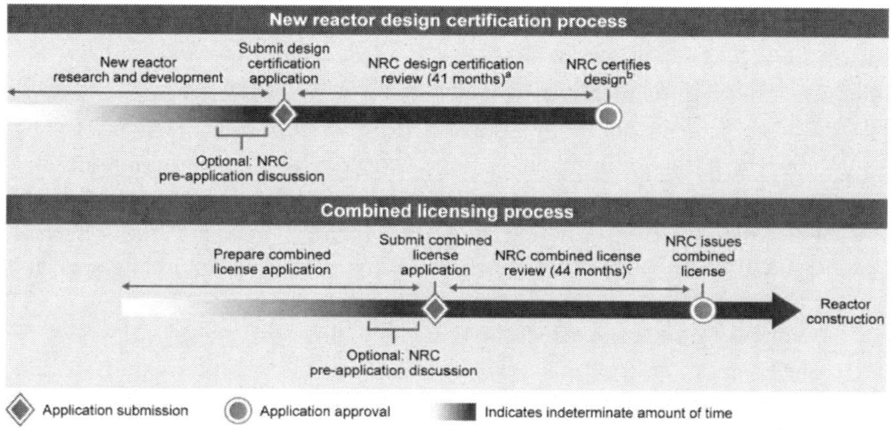

Source: GAO, based on Nuclear Regulatory Commission documentation. | GAO-15-652.

[a] The 41-month review includes a 60 day acceptance review followed by a 39-month design certification application review. The Nuclear Regulatory Commission (NRC) estimates the design certification application review will take 41 months based on the assumption that NRC's requests for additional information from the reactor designers are received and responded to in a timely manner and meet NRC needs.

[b] The certified design may be amended during construction.

[c] The 44-month review includes a 60 day acceptance review followed by a 30-month review for a combined license application that references a certified design and a 12-month period to complete the hearing process. The 30-month review period is increased to 48-60 months for combined license applications that do not reference a certified design. These NRC estimates for the review duration are based on the assumption that NRC's requests for additional information from applicants are received and responded to in a timely manner and meet NRC needs.

Figure 2. Overview of Nuclear Regulatory Commission's (NRC) Part 52 licensing process with notional time frames.

As an example of a way to determine if a nuclear power plant will be competitive, the per MWe cost of generating capacity from the plant is estimated and compared with the cost of alternative energy sources available to meet the demand in the same market. To estimate the costs several metrics are considered. First is the overnight capital cost of a plant—that is, the cost of engineering, procuring, constructing, and licensing the plant, and its associated infrastructure, as if it were paid for instantly and the plant were to be built overnight such that no interest would accrue during its construction. For example, the existing LWRs are generally large and expensive, with a generating capacity around 1000-1600 MWe per reactor at an overnight capital cost of $6 billion or more with construction times of 6 years or more. Second is the financing cost, which includes the investments made by the plant owners and the costs of financing the construction with a loan, which can be significant given the billions of dollars and the long construction time—including potential delays, for example because of licensing uncertainty—associated with building a nuclear power plant. Third are the ongoing operations and maintenance costs for a nuclear power plant, including security and operations staffing and fuel purchasing costs.

A number of issues can affect the costs and competitiveness of the nuclear power plant operations and maintenance. For example, the thermal efficiency of a reactor—how well it converts the heat it generates to electricity—and the simplicity or complexity of the reactor can affect how expensive it is to operate a reactor relative to the electricity it can produce. All other things being equal, a more efficient reactor will generate more electricity for the same operations and maintenance cost, while a simpler reactor can be both less expensive to build and less expensive to maintain, and members of our expert group noted reactors with passive safety systems could provide such simplification. The analysis of whether a reactor will be competitive is complicated by long time frames—besides the multi-decade process to design, license and build a reactor, there is uncertainty in the energy market over the expected 40 to 80 year operational lifetime of the reactor. Therefore, the decision on whether or not to build may not be based solely on profitability, but also on other factors such as the reliability or diversification of power production.

1.4. Nuclear Reactor Operation

Nuclear reactors generate heat by sustaining a fission chain reaction in nuclear fuel. Nuclear fission reactions can occur when a neutron strikes the nucleus of a large atom, causing that nucleus to split, or fission. The result of a fission reaction is typically two fission fragments, or smaller nuclei; two or more new fast-moving neutrons; and significant heat. In a nuclear reactor, the large atoms used for fission are typically the fissile isotopes uranium-235 or plutonium-239.[16] The new neutrons produced by a fission reaction are used to initiate new fission reactions, resulting in a sustained fission chain reaction. The heat generated by this fission reaction is typically used to create steam and drive a steam turbine to generate electricity.

While there are a large number of reactor technologies that can differ significantly, reactor designs generally incorporate certain common components. For example, the fission reaction occurs in the central region of a reactor called the reactor core. The reactor core typically contains the following components:

- **Nuclear fuel.** Nuclear reactors need fissile isotopes, such as uranium-235 and plutonium-239, to sustain chain reactions. Commercial reactors often use uranium that has been slightly enriched in the isotope uranium-235 as their fissile fuel; the rest of the fuel consists of the non-fissile uranium-238, some of which can be converted to plutonium-239 during reactor operation. Some reactors can also utilize thorium-232 to produce uranium-233 for fuel. Uranium-233 is an isotope that, like plutonium-239, raises proliferation concerns because it may be used in nuclear weapons.
- **Fuel Cladding.** In order to hold and contain the nuclear fuel, as well as the fission products that are created during reactor operation, most reactors use fuel cladding to encase the fuel pellets. This cladding may be a zirconium alloy, as it often is in LWRs; stainless steel; silicon carbide; or other materials designed to withstand the extreme conditions of a reactor. Some advanced reactor concepts use liquid fuel, in which case there is no fuel cladding.
- **Moderator.** Thermal reactors use a moderator material to slow down the fission neutrons in order to sustain the fission reaction. In LWRs, the moderator is water, and a certain type of advanced reactor—a high temperature gas cooled reactor—uses graphite as a moderator. Fast reactors are designed to utilize fast neutrons for the fission reactions and fuel breeding or fuel burning and, accordingly, do not use a moderator.
- **Coolant.** To remove heat from the core, a coolant—typically water, a gas, liquid metal, or liquid salt—is circulated through the core. The coolant both prevents the core from overheating (which could damage or melt the fuel, as in the case of the Fukushima accident) and it carries energy, in the form of heat, outside the core. In some reactor types, such as LWRs, the coolant can also function as the reactor's moderator.

- **Reaction control.** Reactors can use different techniques to maintain the fission chain reaction at appropriate rates. For example, control rods, incorporating materials like boron that absorb neutrons to reduce or stop the nuclear chain reaction, may be inserted into reactor cores to provide control over the reaction rate. Neutron-absorbing materials, such as boric acid, may be introduced to the coolant system to achieve a similar effect.

Reactors also have components outside of the reactor core that help transfer heat from the core and create electricity. Certain types of LWRs—pressurized water reactors—contain a pressurizer to help maintain the correct coolant pressure.[17]

The reactor coolant is circulated through the core and is used to generate steam, which then powers a turbine to generate electricity. Some reactor designs may utilize the energy of the coolant in other ways, such as by directly using the heat as process heat in chemical reactions.

Reactors may be classified by size. LWRs that use the same basic technology for the nuclear reactor core can be classified as small, for example the light water SMRs, with an electricity generating capacity of less than 300 MWe, or as large, with capacities of around 1000 MWe or more. See figure 3 for an example large LWR and figure 4 for an example light water SMR. In figure 3, the primary coolant loop uses pumps (not shown) to circulate heated water from the core to the steam generator and back, while in figure 4 this circulation is achieved entirely within the pressure vessel using natural, temperature-driven flow of the water between the reactor elements. An advanced reactor will often utilize similar design concepts with fuel, fuel cladding, coolant, moderators, and reaction control systems, but with modifications due to differences in the types of coolants, fuels, and materials used in the reactor.

Nuclear reactors typically fall into one of two types, based on the neutron spectrum, or neutron energies at which the fission reactions occur:

- **Thermal reactors** optimize the fission reaction rate in their fuel by slowing down, or moderating, the high-energy fast neutrons that are the products of fission reactions, resulting in thermal neutrons. This moderation of the fast neutrons increases the likelihood that a neutron will initiate a fission reaction. Existing large LWRs are thermal reactors.

- **Fast reactors** do not moderate the fission neutrons, instead leaving them fast. Fast neutrons allow these reactors to be more effective than thermal reactors at creating, or breeding, new fuel through neutron bombardment of uranium-238, creating plutonium-239, an isotope that can be used in nuclear weapons.[18]

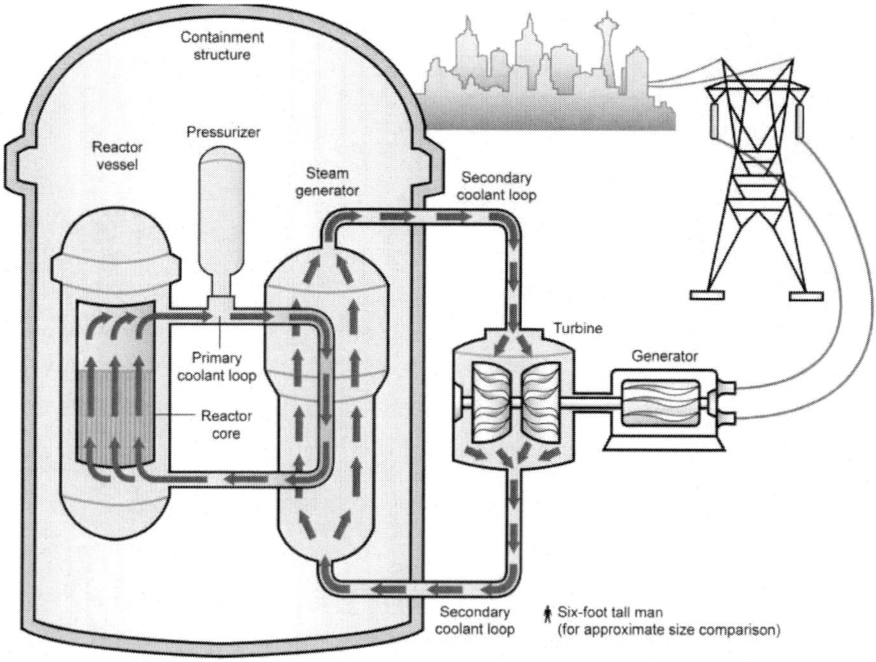

Source: GAO, based on Nuclear Regulatory Commission documentation. | GAO-15-652.

Figure 3. Illustration of a pressurized large light water nuclear reactor (LWR).

Reactors may therefore also be classified by their neutron energies as either thermal or fast reactors, or they may be classified by the materials used in the reactor.[19] For example, an LWR is a thermal reactor using water as both a coolant and moderator, and a gas-cooled fast reactor is a fast reactor using gas (helium) as a coolant with no moderator.

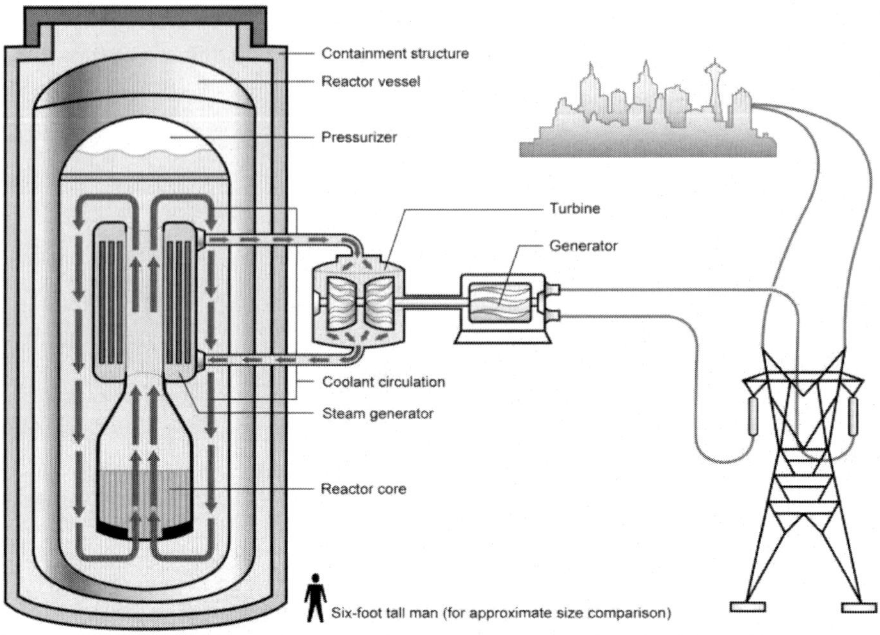

Source: GAO, based on Department of Energy documentation. | GAO-15-652.

Figure 4. Illustration of a light water small modular nuclear reactor (SMR).

2. Four Light Water SMR Designers Have Discussed Certification and Licensing with the NRC, and DOE Has Also Focused on Two Advanced Reactor Concepts

In the United States, four domestic light water SMR designs have been developed to the point that pre-application discussions with NRC have begun, with DOE providing financial support for some of these efforts. Of the four light water SMR designs by reactor designers NuScale, Generation mPower (mPower), Holtec, and Westinghouse, one is developed to the point that a DC application may be submitted to NRC in late 2016, and applications for others may follow, according to the light water SMR designers and DOE officials. DOE has also supported R&D activities on advanced reactor concepts. Specifically, DOE focused support on two advanced reactor concepts that use coolants other than light water—the high temperature gas cooled reactor and the sodium cooled fast reactor—and a third advanced reactor concept using liquid salt receives some support, but these concepts are generally further from certification or licensing and construction than the light water SMR designs. Designs based on these two advanced reactor concepts will likely not have applications submitted to NRC for at least 5 more years, according to DOE and NRC officials.

2.1. One Light Water SMR Design Receiving DOE Support Is Scheduled to Be Submitted for NRC Certification in Late 2016, and Others May Follow

One application for design certification of the light water SMR design under development by NuScale is currently on schedule to be submitted to NRC in late 2016. Time frames have not been set for submission of DC applications for the other light water SMR designs that have been discussed with NRC in pre-application discussions—those by mPower, Holtec, and Westinghouse.

In 2014, DOE entered into a cost-sharing cooperative agreement with NuScale to support certification and licensing efforts, including the DC application, for the NuScale light water SMR design.[20] According to both NuScale representatives and DOE officials we interviewed, as of March 2015 the NuScale design certification application was on track to be submitted to NRC in late 2016. NRC officials estimate that the DC process for this light water SMR will take nearly 3.5 years as a best-case scenario. According to NuScale representatives, this submission schedule could allow their first light water SMR power plant to be operational by 2023 or 2024. NuScale representatives told us they are working with a potential customer and operator for that plant to conduct a site evaluation supporting the COL application that will be needed to license the plant's construction and operation.

In 2013, DOE entered into a similar cost-sharing agreement with mPower, a subsidiary of Babcock & Wilcox, to support the designer's light water SMR certification and licensing efforts. Pre-application discussions between mPower and NRC had resulted in a Design Specific Review Standard, with draft sections issued for public comment through the Federal Register in May 2013. However, mPower decided to scale back efforts on its SMR in mid-2014 because of a lack of committed customers and the need to find additional investors to provide financial support for their certification effort, according to mPower representatives and DOE documentation. DOE continued to provide reduced financial support, in the form of cost-sharing, under this agreement to mPower through November 2014 while mPower attempted to find another investor for the project. When that effort was unsuccessful, DOE suspended further funding and continued the agreement as a no-cost agreement—that is, the agreement is still in place in the event mPower finds an investor, but until that occurs, DOE is not providing funds. As part of the mPower cost-sharing agreement, the Tennessee Valley Authority (TVA) was also receiving funding to support site planning and licensing for an mPower light water SMR power plant at Clinch River, Tennessee. With mPower's reduced licensing effort, TVA representatives told us they have shifted to working on an early site permit application which, if approved by NRC, could allow TVA to build any light water SMR facility that meets the parameters of the permit at that location. As of June 2015, DOE officials told us they were working to finalize an interagency agreement to allow TVA to continue receiving financial support for permitting activity independent of support for mPower.

Holtec and Westinghouse held some pre-application discussions with NRC on their light water SMR designs until 2014. However, while Holtec continues its development work, representatives told us they do not have a detailed schedule for completion of this work, and Westinghouse has suspended its efforts to certify its SMR design. According to Westinghouse representatives, moving forward with significant investments to submit a DC application to NRC was not justified, because they do not have a sufficient number of committed customers in the United States or a DOE cost-sharing agreement. Westinghouse representatives said they may resume efforts to certify their light water SMR design if market conditions change—that is, if they identify a sufficient committed customer base—or if financial risks are otherwise reduced.

In addition to the four light water SMR designs that have been the topics of pre-application discussions with NRC, other SMR designs have been proposed and developed to varying degrees. For example, light water SMR designs are under development or construction internationally, including Argentina, China, France, Russia, and South Korea.

2.2. DOE Has Also Focused R&D Support on Two Advanced Reactor Concepts

DOE has supported R&D for cross-cutting advanced reactor work—work that could be applied to multiple concepts, such as materials studies—as well as development focused on specific advanced reactor concepts. The two specific advanced reactor concepts on which DOE has focused for domestic R&D are the high temperature gas cooled reactor (HTGR) and the sodium cooled fast reactor (SFR). DOE has also supported, to a lesser degree, work on the molten salt reactor (MSR)—specifically, a sub-type of this reactor known as the fluoride salt cooled high temperature reactor (FHR). DOE supports development of these advanced reactor concepts through awards to universities and national laboratories and through cost-sharing arrangements with several reactor designers. All three of these reactor concepts are also being developed internationally, and DOE participates in multi-national efforts, such as the Generation IV International Forum (GIF), an international cooperative endeavor with 13 partners established to carry out R&D for new nuclear energy systems. According to GIF documentation and NRC officials, no advanced reactor is likely to be ready for an application to NRC for at least 5 years.

The HTGR is a high-temperature thermal (graphite-moderated) helium-cooled gas reactor that has a core outlet temperature of 700° to 950° C.[21] The concept is based on commercial gas reactors that have been already built and operated.[22] According to DOE, the high outlet temperature allows the reactor to be used to produce high temperature process heat for use in oil refineries, chemical plants, and the production of hydrogen. This process heat can potentially expand the role that nuclear energy has in energy sectors beyond electricity production by providing an alternative for processes currently using fossil fuels to supply process heat. The outlet temperature also allows for electricity production with thermal efficiencies of 40 to 50 percent, which are high compared with the typical thermal efficiencies for large LWRs of approximately 32 to 34 percent. While HTGR designs would still need safety reviews for licensing or design certification, according to GIF documentation, gas reactors, including HTGRs, are expected to have good safety performance.[23]

The HTGR is regarded by DOE and its international partners as one of the more mature advanced reactor concepts because it is based on gas reactor technology with significant operating experience as well as research progress made over the last 10 years, including work done as part of the Next Generation Nuclear Plant (NGNP) Project.[24] According to GIF documentation, a reactor in China—the 200 MWe HTR-PM— will further demonstrate aspects of the concept's safety and operational feasibility and may be operational in 2017. Additional work on the HTGR continues in areas such as demonstrating the safety features, including passive decay heat removal systems that can prevent the reactor from overheating without external intervention; fuels and materials qualification, including testing of the fuel coatings under postulated accident conditions and the pressure vessel materials for high temperature use; and developing and testing the process heat application concepts.

The SFR uses liquid sodium as a coolant and operates as a fast reactor and thus does not have a moderator. It is a moderately high temperature reactor with an outlet temperature of 500-550° C. While SFR designs would still need safety reviews for licensing or design certification, according to GIF documentation, SFRs are generally expected to have good safety performance; however, the liquid sodium coolant is reactive and may burn if exposed to water or air, requiring that the system be sealed.[25]

DOE and GIF documentation states that the SFR, like the HTGR, is one of the more mature of the advanced reactor concepts, in large part because of experience with similar reactors that operated in the past and with current reactors, such as the operational Chinese Experimental Fast Reactor and Russian BN-600.[26] Additional work is being done to look at safety and severe accident prevention and mitigation, fuel development, and certain component development.

The MSR uses liquid fluoride salt as a coolant. There are generally two types of MSR concepts.

The first dissolves the fissile fuel in the salt coolant itself. The second circulates the salt coolant around the solid fuel pebbles without dissolving the fuel. This second type is sometimes also referred to as a fluoride salt-cooled high-temperature reactor (FHR), and much of the DOE support for MSR development in the United States is focused on the FHR. These reactors operate at atmospheric pressure and high temperature (700°C).

According to members of our expert group and GIF documentation, the MSR/FHR is considered to be a less mature advanced reactor concept than the HTGR or SFR. While two test MSR reactors were operated several decades ago in the United States, there is less experience with operating FHRs.[27] According to GIF documentation, additional work is needed on current commercial MSR concepts, including further studies of the salt chemistry and thermodynamics, as well as the salt interaction with air and water in the event of a severe accident. Components and materials need to be developed and tested for compatibility with the salt, including through corrosion studies. The salt under study for the FHR requires highly enriched lithium-7, which we previously found may face a potential future shortage.[28]

According to GIF documentation, a number of countries, including the United States, have agreed to coordinate on development of six advanced reactor concepts—the three discussed above, on which DOE coordinates work on HTGRs and SFRs through the GIF, as well as the lead-cooled fast reactor, the supercritical water reactor, and the gas fast reactor. According to these documents, DOE has been observing the international efforts on the MSR and lead-cooled fast reactor concepts but has not been participating in those R&D efforts. The lead-cooled fast reactor is primarily under development by Russia, where some small reactors using this technology are under construction, and the work has been informed by lead reactor experiences with the former Soviet naval program. The major source of interest in the supercritical water reactor is in Canada, where it is viewed as a potential successor to their heavy water reactors. However, the supercritical water reactor concept still requires significant R&D with respect to materials and water chemistry. According to documents we reviewed, the gas fast reactor is related to the HTGR but is designed without a moderator to allow it to operate as a fast reactor. This concept requires significant materials and fuel work as well as safety studies, and no gas fast reactor has yet been operated.

3. New Reactor Concepts Could Provide Benefits That May Allow Expansion of Potential Commercial Applications for Nuclear Reactors

Key benefits that light water SMRs or advanced reactors under development are intended to provide include lower construction and financing cost, greater flexibility, and greater operational efficiency. Light water SMR designers intend for their designs to reduce the cost and time of reactor construction and to allow for greater flexibility in the application of nuclear power by providing reactors that can be located in more places than large LWRs and that can be more flexible for meeting various electric grid needs. Advanced reactor designers intend for their designs to provide improved safety, efficiencies, and fuel utilization, as well as increased flexibility to use nuclear reactors for non-electric applications such as supplying process heat or managing spent fuel. However, both light water SMRs and advanced reactor concepts face challenges in development and deployment.

3.1. Light Water SMR Designers Intend for Their Designs to Reduce Cost and Time of Reactor Construction and to Provide Flexibility in Nuclear Power Options

Light water SMRs can provide benefits such as increased flexibility and options for potential reactor operators. Light water SMRs, as their name implies, have two important design features, each of which leads to certain reactor characteristics. First, they are designed to be small compared to large LWRs—both in physical size and in power output. Second, they are designed modularly—that is, components can be manufactured in a factory environment, where standardization and learning of production techniques can help reduce costs, and the components can then be assembled later on-site. Furthermore, light water SMRs are generally designed around the idea that there are trade-offs between the economy of scale that large LWRs provide (by lowering the cost per MWe with very large facilities) and the potential economy of mass production that modular construction of larger numbers of smaller-sized reactors can provide. In February 2015, NuScale announced that their expected overnight construction cost of a light water SMR power plant consisting of twelve reactor modules, with total electric generating capacity of about 570 MWe, would be about $2.9 billion, and that future plants could potentially drop in cost to $2.5 billion—about half of what a single large LWR would cost, and at a similar cost per MWe.[29] However, the operations and maintenance cost estimates for light water SMRs have yet to be fully developed, and while they may be similar to those for existing reactors, several issues related to these costs have yet to be resolved.

Light water SMR characteristics that could provide benefits that increase their flexibility for potential operators include the following:

- **Lower construction cost:** The smaller size of light water SMRs decreases the component costs for the reactor, and the lower cost allows for easier financing. While there is a reduction in power output for a light water SMR as compared to a large LWR, the smaller overall construction cost may make nuclear power, in the form of SMRs, a more viable option for smaller customers—although joint ownership of a large power plant remains an option for them. According to reactor designers, one part of reducing construction costs is simplification of the safety systems, which may also increase overall safety. For example, existing large LWRs require external power or backup generators to run coolant pumps to remove the residual decay heat that is in the reactor core after the reactor is shut down.[30] Light water SMRs can remove heat using fewer pumps and motors or by using passive processes, such as natural circulation of coolant. According to reactor designers, passive safety measures, whether in light water SMRs or in other large LWR designs under construction, can increase the time period that reactors can remain safe— even without external electricity—to several days or longer, possibly indefinitely.[31]

- **Shorter construction times:** While a light water SMR power plant is expected to still require 3 or 4 years to build, large LWRs can take 6 years or more. The shorter construction times for SMRs could allow for easier planning as well as reduced financing costs, and because an SMR power plant may consist of multiple reactor modules, the modules installed first may become operational earlier than the project completion date.
- **Siting flexibility:** Light water SMRs are intended to have a smaller facility footprint, allowing more locations to be considered for a reactor. In addition, due to their smaller fuel loads and design features, according to DOE officials and reactor designers, these light water SMRs may further increase siting flexibility if NRC grants regulatory exceptions to allow them to have smaller exclusion area and emergency planning zones than existing large LWRs. Exclusion areas are based on potential releases of radioactive material and specify maximum population densities around planned reactor sites. The emergency planning zone is the area within about 10 miles of the reactor site for which the operator and state and local entities must prepare predetermined emergency response plans. There is a second emergency planning zone at about 50 miles from a reactor site that requires predetermined plans for monitoring water and food sources.
- **Grid flexibility:** The smaller size of light water SMRs provides potential benefits with respect to the electric grid or distribution systems. First, some markets may not need the 1000 MWe or more that a large LWR provides, so a light water SMR may be better suited for that locality's electricity distribution needs. For example, SMRs may be suitable for replacing coal-fired power plants of similar size. Second, some rural or remote areas can have electricity distribution systems that do not have the capacity to handle the power output from a large LWR.

- **Remote locations:** Some customers in rural or remote areas with high costs associated with shipping fossil fuel to them—and subsequently with high electricity generation costs—may find light water SMRs to be economically competitive. Light water SMRs can provide additional flexibility with electricity generation options for customers in such areas. For example, remote mining operations or isolated villages may find SMRs competitive compared to continuously importing diesel.

3.2. Advanced Reactor Designers Intend for Their Concepts to Provide Higher Efficiencies and Opportunities for Additional Industrial Applications

Advanced reactor designers, including those receiving support from DOE, intend for their concepts to result in reactors that have higher thermal efficiency than either light water SMRs or large LWRs. The thermal efficiency is related to the outlet temperature of the reactor. Large LWRs (as well as light water SMRs) have outlet temperatures around 300° C and a resulting thermal efficiency of around 32 to 34 percent. However, advanced reactors have significantly higher outlet temperatures—500-550° C for SFRs and 700° C or higher for HTGRs and FHRs—allowing them to operate with significantly higher efficiencies of 40 to 50 percent. The higher thermal efficiencies can translate to more economical and competitive reactors. They also have the potential for reducing water consumption requirements for the reactors, according to members of our expert group.

Advanced fast reactors can provide another type of efficiency—they can be designed to consume more of their fuel before needing to be refueled. They can also be operated as breeder reactors, whereby they can create more fuel than they consume, or they can be operated as burners, in which case they can lessen the need for spent fuel storage by consuming some of the spent fuel from LWRs. While uranium prices and reprocessing costs may not change significantly, these options could become increasingly important if uranium fuel costs increase significantly or while uncertainty remains about a permanent spent fuel repository. However, some of these options would require additional R&D and investment in spent fuel reprocessing capabilities. There are also potential nonproliferation considerations. For example, reprocessing facilities are sometimes viewed as proliferation risks because they can provide a means to separate out materials from spent fuel that can be used to make nuclear weapons and would therefore require safeguards and security that could affect the reprocessing operations or economics. Furthermore, the reprocessing facilities still generate waste that must be stored and disposed. According to a reactor designer, one of the advanced reactor concepts—the FHR—may also efficiently store energy in the form of heat, making it a potential option for integration with intermittent renewable energy sources.

According to DOE and reactor designers, advanced reactors—as well as light water SMRs to a lesser degree—are intended to provide a benefit that could potentially allow a new commercial application for nuclear reactors if they were located with industrial processes that could directly use the heat they produce. The relatively high temperatures of advanced reactors could supply process heat for use in oil refineries, chemical plants, and the production of hydrogen—potentially expanding the role that nuclear energy can have in energy sectors beyond electricity production by providing an alternative for processes currently using fossil fuels to supply process heat. For example, at the lower temperature range of water-cooled reactors, including large LWRs and light water SMRs, desalination plants could use heat to distill water for drinking water or agricultural purposes. Some of these industrial applications may also allow the reactors to be used for the industrial application intermittently when electricity demands on the grid are reduced, rather than decreasing the reactors' energy production to match electricity demand. However, according to members of our expert group and NRC documentation, co-locating a nuclear reactor with a potentially hazardous industrial facility will require the license to include the potential impacts of an industrial accident on reactor safety—which is already considered for reactors near other facilities—and it could raise industry concerns that a potentially expensive industrial facility could be forced offline if the reactor shut down or had an accident. DOE explored some of these commercial applications for nuclear-generated process heat as part of its NGNP Project. In 2011, DOE decided not to proceed with the deployment phase of the project, citing several barriers, including being unable to reach an agreement on a cost-share arrangement with industry partners to fund the deployment phase.[32]

Several members of our expert group agreed with the idea that advanced reactors will likely initially be developed with small, modular designs. In this case, in addition to the advanced reactor capabilities described in this section, advanced reactors may also offer the flexibilities of SMRs, as previously discussed.

4. Development and Deployment Efforts for New Reactor Concepts Face Several Challenges

While new reactor concepts may provide some benefits, light water SMRs face some development challenges even though they are similar to existing large LWRs. Advanced reactors, which differ significantly from the existing large LWRs, face more development challenges. Both types of new reactor concepts face some common challenges such as long time frames and high costs associated with the shift from development to deployment—that is, in the construction of the first commercial reactors of a particular type.

4.1. Light Water SMR Development Faces Some Technical, Certification, and Licensing Challenges, and Advanced Reactor Development Faces More Substantial Challenges

While both light water SMRs and advanced reactor concepts are built on knowledge gained over decades of industry and government work in nuclear power, both types of reactors still face some technical, certification, and licensing challenges. Because they use the same coolant and operate under similar temperatures and pressures as the existing large LWRs, as well as recently certified large LWR designs, the four light water SMR designs discussed above share many operational similarities with such reactors, according to DOE and NRC officials. The light water SMRs generally use the same types of reactor concepts as large LWRs, but with the goal of reducing the size and complexity of the reactor. The result is a reactor with similar components and similar operational characteristics, such as water chemistry and temperature, but with a different geometry and, in some cases, components that have been integrated into a single pressure vessel. The light water SMR designs are evolutionary changes to existing large LWRs and, as such, have reduced technical risk. Because of their similarities to reactors widely used in the United States, reactor designers and DOE officials told us they have greater confidence in the performance of the light water SMR designs than in designs using advanced reactor concepts.

According to reactor designers, the work remaining for light water SMRs largely involves the SMR designers finishing the detailed design of the power plants and completing the demonstration of their economics and safety claims, as well as providing support to the NRC during the certification or licensing processes to address claims specific to their intended operation, such as claims that operations and security staff regulatory requirements could be reduced as compared to existing large LWR requirements. If the light water SMR designers are unable to demonstrate that their designs can operate safely without adding to the complexity of the design, their construction and maintenance costs may increase and thus weaken their economic competitiveness. Similarly, if the light water SMR designers are unable to demonstrate that the numbers of per-reactor operations and security staff may be safely reduced from those required for an existing large LWR, the estimated ongoing operations costs may increase and weaken SMR economic competitiveness. According to members of our expert group, light water SMR designers will also need to address some technical challenges related to the manufacturing and assembly of these reactors. Because the light water SMRs rely on standardization of the design of a reactor and producing large numbers of that reactor to lower construction costs, the individual components need to be highly standardized so that they can be assembled properly at the reactor sites. Members of our expert group told us this component standardization has proven challenging for the construction of the Westinghouse AP1000, which is a large LWR that has some modular components, but they also noted light water SMRs have smaller components than the AP1000, so component standardization may be more easily accomplished for SMRs. Despite these remaining challenges, members of our expert group were in nearly unanimous agreement that there were no "show-stoppers" for the NuScale light water SMR design, with one dissenting expert noting that the passive safety features still need to be fully demonstrated.

According to DOE officials, in contrast to the light water SMRs, which are similar to existing large LWRs, advanced reactors face more challenges, in part because the reactor industry has less operating experience with advanced reactors. Advanced reactors also operate at higher temperatures and, for the fast reactors, in a more severe neutron environment. Therefore, according to DOE officials, before advanced reactors can be commercially viable, designers have significantly more R&D issues to resolve, including in areas such as materials studies and fuel certification, coolant chemistry studies, and safety analysis. Some members of our expert group also noted a potential need for new test facilities to support this work. Furthermore, the current NRC certification or licensing processes were described to us by former and current NRC staff as being focused on the reactors that have been built—that is, large LWRs. According to reactor designers, certifying or licensing an advanced reactor may be particularly time-consuming and difficult, and the need for exemptions to and interpretation of the current processes if applied to advanced reactors could introduce economic uncertainty for the applicants.

4.2. New Reactor Development and Deployment May Be Affected by Long Time Frames, High Cost, and Uncertainties

The development of new nuclear reactor designs through the Part 52 process is time consuming, with up to 10 or more years of design work leading to a DC process that NRC projects to take nearly 3.5 years as a best-case scenario, COL applications that can take at least about 4 years if referencing the DC, and construction that can take at least another 3 to 4 years. Members of our expert group, reactor designers, and DOE officials told us that the cost of this process rapidly escalates—early R&D can be done for tens of millions of dollars, while the cost to complete the R&D and to obtain a design certification from the NRC could reach $1 billion to $2 billion. Of this amount, based on the two most recent design certifications NRC has granted, about $50 million to $75 million is for NRC fees related to DC pre-application, application, and revisions, and the rest is spent on reactor R&D and design work. Some reactor designers told us that they have been challenged to find investors for such a costly process when some uncertainties remain about the NRC certification or licensing processes for light water SMRs, and particularly for advanced reactors. Reactor designers seeking customers for their reactors also face challenges arising from uncertainties related to the scale of investment needed to build a new reactor, with general hesitancy among potential customers to commit to a several billion dollar construction project without a demonstration of the technology. According to members of our expert group and reactor designers we spoke with, reactor designers may also have difficulties arising from uncertainties related to the future competitiveness of reactors relative to other forms of electricity production or because of future changes in public perceptions about nuclear power, including concerns about reactor safety and spent fuel disposition. Finally, according to members of our expert group, reactor designers may see a reduced customer market for their new reactor designs if they perceive difficulties or uncertainties with exporting their designs.

New Reactors Can Be Certified or Licensed with Existing Regulations, But Uncertainties Could Increase the Time Needed

According to the NRC, any new reactor technology can be certified or licensed using existing 10 CFR Part 50 or 52 regulations. However, these deterministic regulations were developed for existing large LWRs,[33] so exemptions would be needed for reactor designs differing significantly from existing large LWRs, or the regulations would otherwise need to be adapted, according to reactor designers and NRC officials. According to NRC officials, these exemptions must be specifically applied for by reactor designers or license applicants before the NRC will actively pursue them, and the pre-application discussions between reactor designers and NRC are intended to help identify these exemption items. Several reactor designers told us that they would like regulations changed in order to lessen the uncertainty introduced by relying on exemptions during the DC or licensing process. According to reactor designers, the uncertainty associated with the need for exemptions increases their development risk by potentially increasing the length of the multiple-year DC or license application process. Reducing that uncertainty could help reactor designers find customers—such as utilities—and obtain financing to build plants. A reactor designer representative told us that they were waiting until after the first light water SMR has completed a DC application in order to benefit from a better understanding of the process and potentially updated regulations or exemptions before they would apply for a DC.

Advanced reactors will require more exemptions from current regulations than light water SMRs and thus face greater licensing uncertainties, according to members of our expert group and officials from NRC and DOE. Furthermore, advanced reactors that are designed to produce process heat will need additional scrutiny from NRC because of the need for them to operate safely in proximity to an industrial facility that could affect operations at the reactor site—this NRC requirement applies to any reactors operating near external facilities that may impact safety at the reactor site. For reactors that create process heat for direct use in other applications, such as industrial chemical processes, additional design analysis and regulatory reviews are needed of the interface between the nuclear and non-nuclear process and the degree to which failure of the non-nuclear process could affect safety.

To identify and resolve some of the open policy issues related to advanced reactors and to reduce the licensing uncertainties faced by advanced reactors as a result of the exemptions they would currently need, in July 2013, DOE and NRC established a joint initiative to address portions of the licensing framework they determined to be essential to advanced reactor technologies. In December 2014, the Argonne, Idaho, and Oak Ridge National Laboratories, under a DOE contract, completed the first phase of this work with the production of guidance for developing principal design criteria for advanced, non-light water reactors.[34] NRC officials told us they are evaluating this guidance and anticipate issuing updated regulatory guidance for licensing advanced reactors by December 2016.

First-of-a-Kind Nuclear Plant Costs Increase Construction Challenges

As previously described, nuclear power plant economics involve considerable costs, such as capital costs and plant operating costs. Included in capital costs are site preparation, construction, manufacturing, commissioning and financing a nuclear power plant. Plant operating costs include the costs of fuel, operations and maintenance (O&M), and a contribution to funding the decommissioning of the plant and treating, storing, and disposing of used nuclear fuel and wastes. For newly built nuclear power plants, capital cost recovery is a major driver of the cost of power.[35]

Based, in part, on our review of economic studies for SMRs, the first-of-a-kind (FOAK) plants generally cost more than nth-of-a-kind (NOAK) plants, because FOAK plants are designed to show that a plant is commercially viable and to facilitate the optimization of the construction of a manufacturing plant and supply chain dedicated to NOAK plant production. The overnight capital cost of the FOAK plants would therefore be higher than those of NOAK plants of the same type, which benefit from major design costs having already been expended, a more mature manufacturing and supply chain, and lessons learned from the FOAK plants, so their overnight capital costs decrease. For example, according to members of our expert group, if a new nuclear reactor design is certified by NRC, the cost for the FOAK plant's COL essentially bears the total cost of the DC application, assuming that no further changes to the design are made and no provisions are made to spread out the cost over later NOAK plants. This increased cost for a FOAK plant particularly applies to light water SMRs, and to a greater extent advanced reactors, because they are different enough from existing LWRs that NRC officials told us they will need to certify or license them with exemptions, assuming current regulations remain in place. In fact, some reactor designers told us that the U.S. government should take the risk to certify or license, and build, the first light water SMR, because of the difficulty in getting financing for an unproven design. While SMR designers seek to decrease their NOAK plant costs over time, some studies suggest that existing, large LWRs have not greatly benefitted from industry-wide standardization or learning to date for reasons including intermittent development and production.

In fact, some studies have found that "reverse or negative learning" occurs when increased complexity or operation experience leads to newer safety standards. On a related point, another reactor designer said that the cost and schedule difficulties associated with building the first new design that has been certified by the NRC and started construction in the United States in three decades—the Westinghouse AP1000, a recently designed large LWR—have made it harder for light water SMRs to obtain financing because high-profile problems have made nuclear reactors in general less attractive.[36] However, if costs of the AP1000 units decrease over time through their certified design and modular design, it may show that the nuclear industry could benefit from lessons learned and simplified, modular designs.

In addition to financing the design and certification if Part 52 is used, reactor designers must be able to find a customer to buy, license, and operate the plant. One reactor designer representative told us they would be able to move forward with their design if they had a signed contract with a customer. However, reactor designers also told us customers want a certified design with fewer uncertainties regarding cost before they will commit to buying a plant.

Changes in Alternative Electricity Generation Create Uncertainty for Nuclear Reactor Profitability

The studies and reports we reviewed and members of our expert group that we spoke with indicated that recent decreases in U.S. nuclear energy production may be due to a number of factors, including a decline in natural gas prices and increases in renewable energy generation. Potential changes in future natural gas prices and renewable energy generation capacity create uncertainty in the profitability of nuclear reactors that compete with these energy sources. Regarding natural gas prices, some nuclear plant operators cited price reductions as an important factor in their decisions regarding nuclear power reactor operations. Utilities appear to have increased baseload generation primarily by stepping up production at natural gas-fired combined cycle (NGCC) plants, with the exception of the construction of five additional large LWRs.[37] Utilities may prefer new natural gas power plants as a source of additional baseload and peak capacity because when natural gas prices are relatively low, high-efficiency NGCC power plants can supply electricity at a lower cost than coal-fired generators, and they are faster and less costly to build than nuclear plants.

The declining price of natural gas was a major contributor to the rise in NGCC electricity generation and the decline in coal-fired generation in recent years, and more stringent clean air standards also contributed to the decline in coal generation. NGCC generation capacity can be added to meet forecasted changes in electricity demand by better matching short-term changes in electricity demand, compared to less-flexible, coal-fired generators and large nuclear reactors. In recent years, the number of nuclear plant retirements has increased, in part because of a decline in profitability as low natural gas prices have influenced the relative economics of NGCC and nuclear plants. Since 2012, five nuclear power reactors representing 4.2 GW of capacity have ended power production, and a sixth, the Oyster Creek plant in New Jersey, is expected to shut down in 2019. For example, the Vermont Yankee Nuclear Power Station in Vernon, Vermont, began operations in 1972, and the owners obtained a renewed license in 2011 to operate the plant for an additional 20 years. However, in December 2014, the owners permanently shut down the plant. According to the owners, their decision to shut down the plant was driven in part by lower natural gas prices, which had reduced the comparative profitability of the plant. However, in the future nuclear power could become more economically competitive if factors affecting the costs of producing electricity change. Such changes may include increases in the cost of natural gas or coal, further innovation in nuclear technology that could reduce future costs, increased emphasis on power reliability or diversity of energy sources, or changes in state or federal energy or environmental policy, such as increased requirements to reduce emissions from fossil fuel powered electricity generators or the adoption of a carbon tax.[38]

In addition to low natural gas prices, DOE officials and members of our expert group told us that renewable energy sources, such as wind or solar, can also exert economic pressure on nuclear reactors. Some renewables are intermittent power sources that are prioritized as electricity sources by grid operators because of green energy usage requirements and their near-zero operating costs when that energy source is available. These sources can cause fluctuations in electricity generation that can cause problems for baseload power plants, including nuclear power plants, because they cannot rapidly increase or decrease their power production to meet hourly demands without additional wear on their fuel and equipment. DOE officials and some reactor designers told us they are exploring the use of hybrid energy systems that could develop a way for different power sources to better integrate with each other to allow nuclear power, which could include light water SMRs or advanced reactors, to remain an economic source of electricity.

Changing Perceptions of Nuclear Safety Can Create Uncertainty in the Demand for Nuclear Power

In March 2011, a 9.0-magnitude earthquake and subsequent tsunami devastated northeastern Japan and resulted in equipment failure at the Fukushima Daiichi nuclear power plant. The resulting radiological emergency involved the most extensive release of radioactive material at a nuclear power plant since the 1986 Chernobyl accident. Following this release, the Japanese government evacuated people within 12 miles of the plant, and later extended the evacuation zone to 19 miles. In total, almost 150,000 people were evacuated. In response to the incident, Japan shut down all of its nuclear power reactors, and concerns heightened worldwide about the safety of commercial nuclear power plants. For example, Germany closed 8 of the country's 17 reactors and decided to shut down the remainder by 2022, resulting in more carbon-intensive energy production as the country attempted to use more renewable energy in a baseload capacity manner and compensated for the loss of nuclear generation through the increased use of fossil fuels. In the United States, the Fukushima incident affected some plans to build new nuclear power plants. For example, in 2011, one partner in an ongoing nuclear project in Texas cited multiple uncertainties around nuclear development in the United States related to the Fukushima incident as a reason to stop capital investment in the project.

In response to the Fukushima incident, the federal government established or strengthened a number of standards and requirements related to nuclear energy. For example, NRC accepted 12 recommendations from a task force that NRC had convened in 2011 to review its processes and regulations and determine whether lessons learned from the Fukushima accident could inform its oversight processes. The task force recommended that NRC require licensees to reevaluate and upgrade seismic and flooding protection of reactors and related equipment, strengthen capabilities at all reactors to withstand loss of electrical power, and take other actions to better protect their plants for a low-probability, high-impact event. NRC's activities to strengthen the safety and security of nuclear power plants after the Fukushima incident have increased the costs associated with some existing LWRs, thereby providing a disincentive for nuclear power production. However, these activities could create incentives for new nuclear reactors that will not have costs associated with retrofitting and may be inherently safer. As of October 2014, the two most recent nuclear reactor designs certified by NRC—the Westinghouse AP1000 and the GE Hitachi ESBWR—have more passive safety systems than existing LWR designs. Light water SMRs and advanced nuclear reactors will have to meet or exceed NRC's current safety requirements.

Access to Potential Export Markets May Influence Reactor Designers

In case light water SMRs or advanced reactors are exported, there are non-proliferation safeguards in some of the new reactors. According to DOE officials and reactor designers, some advanced reactors use fuel encapsulated in materials such as silicon carbide, which make it difficult to access the fuel, potentially making the fuel safer in the event of an accident and more proliferation resistant. Some advanced SMR designers go a step further and claim that their designs could be "black boxed" (that is, they could be deployed already fueled and sealed), and once the fuel is spent, the entire unit could be shipped back to the factory for waste handling and reprocessing. If responsibility for the fuel cycle were thus taken out of the hands of the reactor operator, then risks of proliferation could potentially be reduced. DOE officials told us that these safeguards make such reactors attractive for export purposes. However, the details of how to provide assurance that such reactors remain sealed or how to determine if there has been an attempt to open the reactor have yet to be resolved.

Reactor designers have mixed opinions regarding the feasibility of obtaining export authorization, but most agree that more could be done to ease the process. In addition, we previously recommended that DOE improve its export control process, including efficiency in reviewing applications and found that U.S. designers were at a disadvantage for exporting nuclear designs compared to foreign-state owned or supported nuclear industry.[39] However, some reactor designers are still pursuing international markets, particularly in China, for constructing their reactors, because higher fossil fuel prices in other countries may make their designs more economically competitive. Further, by first demonstrating the construction and operation of a reactor in another country, designers may reduce challenges associated with certification or licensing and construction of their designs in the U.S. market.

5. Concluding Observations

While commercial nuclear reactors provide nearly 20 percent of the electricity in the United States, the existing large LWR fleet is aging while electricity demand in the coming decades may increase. Five large LWRs are currently under construction, but reactor designers have also been working on small modular reactors and advanced reactors that can provide different sets of capabilities than large LWRs. One light water SMR designer may begin its DC application with NRC in late 2016 with a planned 12-reactor power plant completed as early as 2023, while advanced reactors are likely at least 5 years from submitting a design certification application.

New SMR designs and advanced reactor concepts are intended to provide certain benefits. Light water SMRs could expand commercial applications for nuclear power by offering power plants that are lower in construction and financing cost—although with smaller generating capacities—and that have shorter construction times. These smaller reactors can provide more flexibility in siting options and with rural or remote grid locations. Advanced reactors can achieve higher temperatures than LWRs and thus can be used directly for certain industrial applications that currently rely on fossil fuels to achieve high temperatures. Certain advanced reactors, such as fast reactors, may also provide additional fuel cycle flexibility and nuclear waste management capabilities.

While light water SMRs and advanced reactors may provide some benefits, their development and deployment face a number of challenges. Both SMRs and advanced reactors require additional technical and engineering work to demonstrate reactor safety and economics, although light water SMRs generally face fewer technical challenges than advanced reactors because of their similarities to the existing large LWR reactors. Depending on how they are resolved, these technical challenges may result in higher-cost reactors than anticipated, making them less competitive with large LWRs or power plants using other fuels. However, nuclear reactors, including these new reactors, may still be attractive for their reliability, zero carbon emissions, and as a means to diversify energy sources. In the future, nuclear power could become more economically competitive with other energy sources if factors affecting the costs of producing electricity change. Such changes may include increases in the cost of natural gas or coal, further innovation in nuclear technology that could reduce future costs, or changes in state or federal energy or environmental policy, such as increased requirements to reduce emissions from fossil fuel powered electricity generators or the adoption of a carbon tax.

Both light water SMRs and advanced reactors face additional challenges related to the time, cost, and uncertainty associated with developing, certifying or licensing, and deploying new reactor technology, with advanced reactor designs generally facing greater challenges than light water SMR designs. It is a multi-decade process, with costs up to $1 billion to $2 billion, to design and certify or license the reactor design, and there is an additional construction cost of several billion dollars more per power plant. Furthermore, the licensing process can have uncertainties associated with it, particularly for advanced reactor designs. A reactor designer would need to obtain investors or otherwise commit to this development cost years in advance of when the reactor design would be certified or available for licensing and construction, making demand (and customers) for the reactor uncertain. For example, the price of competing power production facilities may make a nuclear plant unattractive without favorable rates set by a public authority or long term prior purchase agreements, and accidents such as Fukushima as well as the ongoing need for a long-term solution for spent nuclear fuel may affect the public perception of reactor safety. These challenges will need to be addressed if the capabilities and diversification of energy sources that light water SMRs and advanced reactors can provide are to be realized.

6. Agency and Third Party Comments and Our Evaluation

We provided a draft of this report to DOE for review and comment. DOE did not provide a written response, but did provide technical comments that we incorporated as appropriate throughout the report.

We provided a draft of this report to NRC for review and comment. NRC provided technical comments that we incorporated as appropriate throughout the report.

We provided a draft of this report to 18 members of our expert group for review and comment and received comments from 17 of them. The majority of the comments were generally favorable. These comments included technical corrections and suggestions that we incorporated throughout the draft as appropriate.

A couple members of our expert group indicated that the report would benefit from more balance with increased input from utility representatives and environmental groups. We did gather information from representatives of some of these groups over the course of this work or related work, and we believe that this report is appropriately balanced with respect to the stated objectives. A couple members of our expert group noted that the report did not directly address structural issues that affect nuclear reactor development, including the need for test facilities or the implications of government support for this work. One member of our expert group suggested a review of the last several decades of government-supported advanced reactor development, and one member of our expert group suggested a more general discussion of the importance of a robust U.S. nuclear industry. While these are all relevant to the topics addressed by this report, we limited the scope of this work to the stated objectives and the supporting information needed for general context.

Several members of our expert group suggested that we include a discussion of the effect of subsidies for renewable energy or fossil fuel sources on the competitiveness of nuclear energy or otherwise include a more comprehensive discussion of the economic role of nuclear energy, including factors such as the prioritized use of renewables over other sources of electricity, improved energy efficiencies, or improved energy storage technologies. One member of our expert group identified reliability issues with some advanced reactors that had operated in the past and questioned the economic viability of reactors, particularly fast reactors operating as burners for fuel cycle flexibility. While these topics are relevant to some of the stated objectives of this report, we determined that a full, balanced study of energy subsidies, the energy market, and the highly uncertain nature of future energy markets, was beyond the scope of this work.

Several members of our expert group commented that the NRC regulatory process was insufficient for licensing new types of reactors—particularly non-light water reactors, such as advanced reactors—without introducing excessive risk for the reactor designer, for example through the necessary use of exemptions. One member of our expert group suggested we provide a comparison of the NRC process to other regulatory structures in the United States and noted the NRC process encourages delays in investment in reactor designs until one designer acts as a first mover and resolves perceived licensing uncertainties. Some members of our expert group suggested changes in the licensing process to reduce this uncertainty, for example by using phased licensing, provisional licensing, or otherwise encouraging NRC to set requirements earlier in the process. Other members of our expert group disagreed and told us such criticism of NRC is largely misplaced, for example by stating the licensing uncertainties are not due to the process but rather to uncertainties in the safety of the designs themselves or a lack of operating experience, and pointing out that if NRC sets requirements earlier in the process, NRC will have less flexibility later in the process when more is known about the reactor design. This report discusses the potential challenges related to licensing that reactor designers perceive as well as steps NRC and DOE are taking to look at the licensing process, and a comparison of the NRC regulatory processes to those of other U.S. regulatory agencies is outside the scope of this work.

Sincerely yours,

Timothy M. Persons, Ph.D.
Chief Scientist

Frank Rusco
Director, National Resources and Environment

APPENDIX I: EXPERT PARTICIPATION

At our request, the following individuals helped to provide information about nuclear reactor development, including reactor economics, reactor development, and licensing. They also reviewed and provided comments, which were incorporated as appropriate, on the draft of this report.

Margot Anderson
Executive Director, Energy Project
Bipartisan Policy Center

George Apostolakis
Former Commissioner
U.S. Nuclear Regulatory Commission

Greg Ashley
President
Bechtel Nuclear Business Line

Douglas Chapin
Principal
MPR Associates, Inc.

Thomas Cochran
Consultant and Retired Senior Scientist
Natural Resources Defense Council

Ashley Finan
Senior Project Manager
Clean Air Task Force

Charles Forsberg
Director, High Temperature
Salt-Cooled Reactor Project
Massachusetts Institute of Technology

Philip Hildebrandt
Special Assistant to Laboratory Director
Idaho National Laboratory

Andrew Kadak
President
Kadak Associates, Inc.

John Kelly
Deputy Assistant Secretary
for Nuclear Reactor Technologies
U.S. Department of Energy

Jim Kinsey
Regulatory Affairs Director
Idaho National Laboratory

Edwin Lyman
Senior Scientist
Global Security Program
Union of Concerned Scientists

William Madia
Vice President for
SLAC National Accelerator Laboratory
Stanford University

David Matthews
Director New Reactor Licensing (retired)
U.S. Nuclear Regulatory Commission

Regis Matzie
Senior Vice President and
Chief Technical Officer (retired)
Westinghouse Electric

Per Peterson
Professor
University of California at Berkeley

Harold Ray
Executive Vice President (retired)
Southern California Edison

Everett Redmond
Senior Director for Policy Development
Nuclear Energy Institute

Jose Reyes
Chief Technology Officer
NuScale Power

Kevan Weaver
Director of Technology Integration
TerraPower, LLC.

End Notes

[1] Nuclear proliferation is the spread of nuclear weapons, fissile materials, and weapons-applicable nuclear technology and information to nations not recognized by the Treaty on the Nonproliferation of Nuclear Weapons.

[2] A light water reactor refers to a reactor of any size that uses light water (ordinary water) to cool the reactor, as opposed to heavy water which contains deuterium, an isotope of hydrogen. Throughout this report, large LWR refers to the existing fleet of large light water commercial nuclear reactors.

[3] A power plant's electricity generating capacity is measured in megawatts of electricity, or MWe. A power plant also can be measured by the thermal energy (heat) it produces, megawatt-thermal (MWth). The ratio of MWe to MWth is the thermal efficiency of the reactor, a measure of how well it converts heat to electricity.

[4] Throughout this report, we refer to light water SMRs, but advanced reactor technologies using other coolants are also being developed as SMRs.

[5] GAO, *Advanced Reactor Research: DOE Supports Multiple Technologies, but Actions Needed to Ensure a Prototype is Built*, GAO-14-545 (Washington, D.C.: June 23, 2014). Advanced reactors, also sometimes referred to as Generation IV reactors, generally use coolant other than ordinary water. While some advanced reactors have been studied and operated in the past, current DOE work on advanced reactors seeks to support commercialization of these new reactors.

[6] The two most recent design certifications—for the Westinghouse AP1000 and the GE Hitachi ESBWR—cost about $45 million and $69 million, respectively, for NRC fees related to pre-application, application, and revision reviews.

[7] NRC estimates design certification for a new LWR design will take 41 months, including a 60 day acceptance period, as a best-case scenario.

[8] In this report, we use the term "customer" to refer to any electricity generator that would use a nuclear reactor to produce electricity, such as an investor or municipal owned utility, or to generate process heat, such as an industrial plant.

[9] There are also intermediate load plants that are used during the transition between baseload and peak load needs.

[10] Some power plants using natural gas have been used as baseload plants. For additional information, see GAO, *Electricity: Generation Mix Has Shifted, and Growth in Consumption Has Slowed, Affecting System Operations and Prices*, GAO-15-524 (Washington, D.C.: May 29, 2015).

[11] For additional information, see GAO-15-524 and GAO, *Energy Policy: Information on Federal and Other Factors Influencing U.S. Energy Production and Consumption from 2000 through 2013*, GAO-14-836 (Washington, D.C.: September 30, 2014).

[12] EIA is the statistical agency within DOE that collects, analyzes, and disseminates independent information on energy issues. EIA notes that the AEO2015 projections are based generally on federal, state, and local laws and regulations in effect at the end of October 2014. The projections generally do not reflect the potential impacts of pending or proposed legislation, regulations, and standards—or of sections of existing legislation that require implementing regulations or funds that have not been appropriated. In certain situations, however, where it is clear that a law or a regulation will take effect shortly after AEO2015 is completed, that law or regulation may be considered in the projection.

The AEO2015 Reference case projection is a business-as-usual trend estimate, given known technology and technological and demographic trends. Energy market projections are subject to much uncertainty. Many of the events that shape energy markets are random and cannot be anticipated, such as the effect that heavy rainfall can have on hydroelectric power and the resulting reduction in electricity prices. In addition, future developments in technologies, demographics, and resources cannot be foreseen with certainty.

[13] The DC applications were originally approved by the NRC in about 4 years for the AP1000 and 9 years for the ESBWR. Subsequent revisions to the reactor designs have increased the review period.

[14] The NRC is required to recover approximately 90 percent of its annual budget authority through fees charged to applicants and licensees. Under 10 CFR Part 170, applicants for a construction permit or operating license under 10 CFR Part 50; an early site permit, standard design certification, standard design approval, manufacturing license, or combined license under 10 CFR Part 52; and requests related to pre-application consultations and reviews from the NRC will be assessed a fee to recover NRC's costs for providing those services.

[15] During the last two decades, some state governments and the federal government have taken steps to restructure the wholesale electricity markets with the goal of increasing competition and prices are now largely determined by the interaction of supply and demand rather than regulatory bodies. The electricity industry has historically been characterized by utilities that were integrated and provided the four functions of electricity service—generation, transmission, distribution, and system operations—to all retail consumers in a specified area. In much of the Western, Central, and Southeastern United States, retail electricity delivery continues to operate under this regulatory approach, and these regions are referred to as traditionally regulated regions. In parts of the country where states have taken steps to restructure retail electricity markets, utilities compete with other qualified providers who may not own generation, transmission, or distribution assets to provide electricity to retail consumers by offering electricity plans with differing prices, terms, and incentives.

[16] Isotopes are varieties of a given chemical element with the same number of protons but different numbers of neutrons. For example, the helium-3 isotope, which is used in research and in radiation detection equipment, has one less neutron than the helium-4 isotope, which is the helium isotope commonly used in party balloons.

[17] Another type of LWR—the boiling water reactor—is also used in the United States. Approximately one-third of the operating commercial nuclear reactors in the United States are boiling water reactors.

[18] Although thermal reactors produce plutonium-239, because fast reactors can be particularly effective at fissioning plutonium-239 and changing some isotopes into fissionable isotopes, fast reactors are more likely to raise proliferation concerns. This is because their fuel cycle, using plutonium-239, can involve fuel reprocessing facilities that may pose proliferation risks. Fast reactors optimized for fuel production are called fast breeder reactors and can produce more fuel through breeding than they consume. Because fast reactors may use reprocessed spent fuel from other nuclear reactors as fuel they may reduce the need for long-term disposal of spent fuel.

[19] Some reactor technologies are "epithermal" and fall in between thermal and fast reactors.

[20] This cost share agreement provides matching support from DOE for NuScale costs associated with licensing the NuScale design with NRC. The DOE support can be up to $217 million over 5 years.

[21] HTGRs designed for the higher end of these temperatures may also sometimes be referred to as VHTRs – very high temperature reactors.

[22] These gas reactors include the Fort Saint Vrain and Peach Bottom 1 reactors, commercially operating 1979-1989 and 1967-1974, respectively, in the United States.

[23] The expected HTGR safety performance is based on a number of factors, described below, that together make it difficult for an accident to cause the HTGR core to increase in temperature to a sufficient degree that fuel and fission products can be released from the reactor. HTGRs have a strong negative temperature coefficient of reactivity, meaning that as temperatures increase the nuclear reaction slows down. In addition, the graphite moderator has a high heat capacity, meaning it will heat up relatively slowly, and the core has a low power density, meaning the reactor is better able to passively remove decay heat. Moreover, the nuclear fuel currently in development for use in the HTGR consists of many small spherical kernels of fuel, each individually coated with materials that allow very high temperatures to be reached before the fuel and its fission products will be released, and the reactor vessel relies on natural heat transfer to ensure fuel temperatures remain well below potential fuel failure points in the event coolant is lost.

[24] NGNP was established by the Energy Policy Act of 2005. Under the act, DOE is to deploy a prototype NGNP reactor using advanced technology to generate electricity, produce hydrogen, or both, by the end of fiscal year 2021. However, in 2011, DOE decided not to proceed with the deployment phase of the project, citing several barriers. For additional information see GAO-14-545.

[25] The expected SFR safety performance is based on a number of factors. The coolant has a high thermal inertia, meaning it is harder to increase the temperature of the coolant, and it has a good temperature margin before boiling—that is, the reactor operates at coolant temperatures well below where the coolant might begin to boil. The coolant can be held at atmospheric pressure, reducing the likelihood of a loss of coolant accident.

[26] Other examples of SFRs are, among others, the Fast Flux Test Facility, operated in the United States from 1980 to 1992; the Experimental Breeder Reactor-II, operated in the United States from 1964 to 1994; the Fermi reactor, operated in the United States from 1963 to 1975; and the Phénix reactor, operated in France from 1973 to 2009.

[27] These two salt-cooled reactors operated in the United States were the Aircraft Reactor Experiment in 1954 and the Molten-Salt Reactor Experiment from 1965 to 1969.

[28] GAO, *Managing Critical Isotopes: Stewardship of Lithium-7 Is Needed to Ensure a Stable Supply*, GAO-13-716 (Washington, D.C.: September 19, 2013).

[29] According to NuScale representatives, the plant cost estimate for the first plant assumes a generic location in the southeast region of the United States, does not include owner costs such as licensing or transmission interconnect, and has an expected accuracy of +35%/-10%. According to NuScale representatives, the future plant cost estimate is less firmly supported than the estimate for the first plant.

[30] The failure of backup power supplies was a contributing factor to the Fukushima accident in 2011.

[31] The Westinghouse AP1000, a large LWR, has a design certified by NRC and uses some of these passive safety measures. Four AP1000 reactors are under construction in Georgia and South Carolina.

[32] See GAO-14-545.

[33] Deterministic regulations are not based on numerical estimates of risk, but rather on experience, test results, expert judgment, engineering margins, and the concept of defense-in-depth; for example, current NRC regulations require that no more than two reactor units may be operated from a single control room, although at least one light water SMR designer hopes to operate more than two units from a common control room. NRC regulations have been shifting from a deterministic to "risk-informed" and/or "performance-based" approach for certain topics, but deterministic regulations are still present and are largely based on the experiences with existing large LWRs.

[34] Idaho National Laboratory, *Guidance for Developing Principal Design Criteria for Advanced (Non-Light Water) Reactors* (Idaho Falls, Idaho: December 2014).

[35] Once capital costs are paid off, plant operating costs become the major components of the cost of power.

[36] The AP1000 was the first new design that has been certified by the NRC and started construction in the United States in three decades. However, construction problems, including supply chain and regulatory issues, have resulted in cost and schedule increases. For example, for the AP1000 units under construction at the V.C. Summer Nuclear Station in South Carolina, NRC inspectors determined that the rebar spacing and depth was not in alignment with the design certification document, which required reanalysis and resulted in

about a six month delay. Quality control issues with key submodules for this same reactor have also caused a schedule delay and cost increases.

[37] There are four reactor units—two at each location—under construction at the Vogtle Plant in Georgia and V.C. Summer Nuclear Station in South Carolina. There is one additional unit under construction at the Watts Barr Nuclear Power Plant in Tennessee.

[38] For example, an executive order issued on March 19, 2015 regarding federal leadership in sustainability and greenhouse gas reductions specifically noted using small modular reactors among other alternative energy options as a way to meet clean energy targets for federal agencies.

[39] GAO, *Nuclear Commerce: Governmentwide Strategy Could Help Increase Commercial Benefits from U.S. Nuclear Cooperation Agreements with Other Countries*, GAO-11-36 (Washington, D.C.: November 4, 2010), and *Nuclear Commerce: Additional Actions Needed to Improve DOE's Export Control Process*, GAO-15-124 (Washington, D.C.: October 14, 2014).

In: New Nukes
Editor: Darnell Walters

ISBN: 978-1-63484-553-3

Chapter 2

ADVANCED REACTOR RESEARCH: DOE SUPPORTS MULTIPLE TECHNOLOGIES, BUT ACTIONS NEEDED TO ENSURE A PROTOTYPE IS BUILT*

United States Government Accountability Office

WHY GAO DID THIS STUDY

NE conducts R&D on advanced nuclear reactor technologies with multiple aims, including (1) improving the economic competitiveness of nuclear technology to ensure that nuclear power continues to play a role in meeting our nation's energy needs; (2) increasing safety; (3) minimizing the risk of nuclear proliferation and terrorism; and (4) addressing environmental challenges, such as reducing greenhouse gas emissions. External groups have been critical of NE for, among other things, how it prioritizes advanced reactor R&D.

GAO was asked to review NE's advanced reactor R&D efforts. This report (1) describes NE's approach to advanced nuclear reactor R&D and (2) examines how NE plans and prioritizes its advanced reactor R&D activities, including deploying an advanced reactor. GAO reviewed laws and reports concerning NE's efforts to develop advanced reactor technologies and

* This is an edited, reformatted and augmented version of the United States Government Accountability Office publication, GAO-14-545, dated June 2014.

interviewed NE officials and a nonprobability sample of companies developing such technology, selected because of their involvement with DOE's R&D efforts.

WHAT GAO RECOMMENDS

To better prepare DOE to meet the requirement of EPAct 2005 to deploy the NGNP prototype reactor, GAO recommends that DOE develop a strategy for resuming the NGNP Project and provide a report to Congress updating the status of the project. DOE agreed in principle with GAO's first recommendation and respectfully disagreed with the second. GAO believes these recommendations remain valid as discussed in the report.

WHAT GAO FOUND

The Department of Energy's (DOE) Office of Nuclear Energy's (NE) approach to advanced reactor research and development (R&D) focuses on three reactor technologies—high-temperature gas-cooled reactors, sodium-cooled fast reactors, and fluoride-salt-cooled high-temperature reactors—but NE is also funding research into other advanced reactor technologies. NE's approach is to conduct research in support of multiple advanced reactor technologies, while collaborating with industry and academia, with the ultimate goal for industry to take the results of NE's research to the next step of development and commercialization. This approach provides several advantages, including flexibility in responding to changes in future U.S. energy policy. Many representatives that GAO talked to from the nuclear power industry and the National Academy of Sciences agree with NE's approach, saying that current policies on controlling greenhouse gas emissions and disposing of nuclear waste do not make a compelling case for choosing a reactor technology to develop. However, others GAO talked to are critical of some of the reactor technologies NE chooses to research, citing economic and technological challenges. The Nuclear Energy Advisory Committee has criticized NE's approach, recommending that NE focus its efforts on a smaller number of technologies to help ensure that a reactor prototype is deployed. To remain aware of industry's R&D needs and international nuclear energy developments, NE regularly collaborates with industry and international organizations.

NE uses internal and external reviews to set program and funding priorities for advanced reactor R&D activities and to evaluate progress toward program goals. For example, NE conducts internal monthly and quarterly reviews to discuss project status, budgets, and technical highlights. Furthermore, NE's R&D efforts are periodically reviewed by external entities, including the Nuclear Energy Advisory Committee. Among the advanced reactor technologies that NE's R&D currently supports, the high-temperature gas-cooled reactor is the technology that is most likely to be deployed and commercialized in the near term, according to an NE planning document. NE officials said this likelihood is based on the wide range of potential industry market applications and because of substantial government investments in the technology's development. NE has been pursuing this technology under the Next Generation Nuclear Plant (NGNP) Project, as established by the Energy Policy Act of 2005 (EPAct 2005). Under EPAct 2005, DOE is to deploy a prototype reactor for NGNP by the end of fiscal year 2021. However, in 2011, DOE decided not to proceed with the deployment phase of this project, citing several barriers. For example, NE and industry have been unable to reach an agreement on a cost-share arrangement to fund the deployment phase because of a disagreement on the applicable cost-share levels and how and when the cost-share would be applied to specific activities or project phases. Although NE continues to conduct R&D for the NGNP Project, it has not developed a strategy to overcome the cost-share issue and other barriers to resuming the deployment phase of the project. Furthermore, DOE has not selected initial reactor design parameters or reported to Congress on an alternative date for making this selection. Without doing so, it is not clear when NE is going to take this next step in deploying the NGNP prototype reactor and it risks the project not being completed by the targeted date in 2021.

ABBREVIATIONS

DOE	Department of Energy
EPAct 2005	Energy Policy Act of 2005
FACA	Federal Advisory Committee Act
NE	Office of Nuclear Energy
NGNP	Next Generation Nuclear Plant
NRC	Nuclear Regulatory Commission

OECD	Organisation for Economic Co-operation and Development
R&D	research and development

***

June 23, 2014

The Honorable Dianne Feinstein
Chairman
Subcommittee on Energy and Water Development
Committee on Appropriations
United States Senate

Dear Madam Chairman:

Energy demand in the United States is expected to continue to grow over the coming decades, according to the U.S. Energy Information Administration, and energy security and greenhouse gas emissions from the burning of fossil fuels will continue to be of serious concern. Nuclear energy accounts for about 20 percent of electricity generation in the United States and produces no air pollution or greenhouse gases. However, the United States faces several challenges in developing additional nuclear energy facilities. For example, the March 2011 accident at Japan's Fukushima Daiichi commercial nuclear power plant has increased concerns about the safe operation of nuclear power plants worldwide. In addition, many uncertainties surround long-term storage and disposal options for spent nuclear fuel from existing power plants.[1] Spent nuclear fuel is considered one of the most hazardous substances on earth, and exposure can cause environmental harm and long-term health hazards, such as cancer, in those who are exposed to it. Spent nuclear fuel also raises concerns over nuclear proliferation and terrorism[2] Specifically, concerns exist that terrorist networks could divert nuclear material intended for peaceful purposes to the development of nuclear weapons. In addition, the cost of developing, deploying, and operating new nuclear power facilities in the United States, relative to other types of energy sources, has been an impediment to increasing nuclear production.

Research on a new generation of advanced nuclear reactors, which are intended to improve on the current generation of nuclear reactors through increased safety, improved economic competitiveness, and reduced nuclear waste, is being conducted by governments and private industry around the world. In the United States, the Department of Energy (DOE) spends more than $80 million per year to develop advanced nuclear reactors with the aim of eventually replacing aging U.S. commercial light water reactors,[3] many of which are operating past the end of their initial 40-year operating period under renewed operating licenses.[4] DOE's Office of Nuclear Energy (NE) conducts research and development (R&D) on advanced reactors with multiple aims, including (1) improving the economic competitiveness of nuclear energy to ensure that nuclear energy continues to play a role in meeting our nation's energy needs, (2) increasing safety, (3) minimizing the risks of nuclear proliferation and terrorism, and (4) addressing environmental challenges such as reducing greenhouse gas emissions.[5]

In September 2006, we reported on NE's progress in deploying an advanced reactor prototype,[6] stating that initial R&D results were favorable, but DOE officials considered the schedule to be challenging, given the amount of R&D that remained to be conducted. External groups, such as the Nuclear Energy Advisory Committee,[7] the National Academy of Sciences, and we have reported on NE's advanced nuclear reactor R&D efforts from 2008 to 2013.[8] These reports have been critical of how NE prioritizes its advanced reactor R&D, the advanced reactor technologies that NE has chosen to fund, and how NE partners with industry to advance program goals

In this context, you asked us to review NE's advanced reactor R&D efforts. Our objectives were to (1) describe NE's approach to advanced reactor R&D and (2) examine how NE plans and prioritizes its advanced reactor R&D activities, including deploying an advanced reactor prototype.

To address these objectives, we reviewed relevant federal laws and regulations, as well as reports by DOE, us, and others that describe NE's efforts to develop advanced reactor technologies. In addition, to better understand NE's approach to and decision-making processes for advanced reactor R&D, we interviewed NE officials; representatives from a nonprobability sample of companies developing advanced reactor technology;[9] knowledgeable members of the Nuclear Energy Advisory Committee and the National Academy of Sciences' National Research Council; and representatives from industry groups, including the Next Generation Nuclear Plant Industry Alliance and the Nuclear Energy Institute.[10] In addition, we visited two of DOE's national laboratories—Oak Ridge National Laboratory and Idaho National Laboratory—to better understand NE's R&D activities for advanced nuclear reactors. To select these laboratories, we analyzed the appropriated funding levels associated with NE across the 10 national laboratories that NE funds.[11] We selected the 2 laboratories with the highest levels of funding dedicated to NE programs over the past 5 years.

We conducted this performance audit from August 2013 to June 2014 in accordance with generally accepted government auditing standards. Those standards require that we plan and perform the audit to obtain sufficient, appropriate evidence to provide a reasonable basis for our findings and conclusions based on our audit objectives. We believe that the evidence obtained provides a reasonable basis for our findings and conclusions based on our audit objectives.

BACKGROUND

According to its 2010 *Nuclear Energy Research and Development Roadmap: A Report to Congress*, NE's primary mission is to advance nuclear power as a resource capable of meeting the nation's energy supply, environmental, and energy security needs by resolving technical, cost, safety, proliferation resistance, and security barriers through research, development, and demonstration, as appropriate. NE conducts research aimed at (1) improving the reliability, sustaining the safety, and extending the operational lifetime of existing light water nuclear reactors; (2) supporting the development of the next generation of nuclear reactors, including light-water-reactor-based small modular reactors and advanced reactors, with a focus on affordability; (3) developing sustainable nuclear fuel cycles; and (4) reducing the risk of nuclear proliferation and terrorism. Light-water-reactor-based small modular reactors are smaller in size and energy output than conventional light water reactors—but use the same basic technology for the nuclear reactor core—and offer several potential advantages over existing light water reactors, including lower capital and construction costs through factory fabrication; enhanced safety and security; improved operation times and longer life cycles; and flexibility to be sited at locations that cannot support large nuclear plants, such as isolated areas or sites with limited water supplies.

Advanced reactors, including advanced small modular reactors, use innovative nuclear fuels, coolants, and energy systems, and offer the potential for significant advantages over existing light water reactors, including greater energy conversion efficiency, reduced plant size, lower construction and operation costs, and improved safety. NE has mainly conducted advanced reactor R&D on high-temperature reactors and fast reactors. High-temperature reactors produce electricity, as well as process heat that can be used for industrial purposes, such as refining petroleum or producing hydrogen, and replace current sources of process heat from burning natural gas or other fossil fuels, which emit greenhouse gases. Fast reactors can use spent nuclear fuel as their fuel source, which reduces the need for long-term storage of spent nuclear fuel, and would more efficiently use uranium, helping reduce nuclear waste.

NE conducts nuclear reactor R&D through its Reactor Concepts Research, Development, and Demonstration program, which aims to (1) help advance nuclear power as a resource capable of meeting the nation's energy, environmental, and national security needs and (2) develop new and advanced reactor designs and technologies that advance the state of reactor technology and improve the economic competitiveness of nuclear power. The program encompasses the following subprograms:

- The Light Water Reactor Sustainability subprogram is developing the scientific basis to extend existing nuclear power plant operating life beyond the current licensing period and ensure their long-term reliability, productivity, safety, and security. This subprogram conducts research into materials aging and degradation, updating instrumentation and controls, and assessing reactor safety margins, among other things.
- The Advanced Reactor Concepts subprogram[12] supports the development of innovative reactor technologies that may offer improved safety, functionality and affordability; more efficient energy conversion; increased proliferation resistance and security; and that build upon existing nuclear technology and operating experience. It supports research to reduce technical barriers for advanced nuclear energy systems, and its efforts support various reactor technologies at different maturity levels.
- The Advanced Small Modular Reactor subprogram[13] supports the development of innovative small modular reactor designs that potentially offer improved safety, functionality, and affordability; more efficient energy conversion; increased proliferation resistance and security; and simplified operation and maintenance. For example, the program supports research into novel sensors and control systems for multiple reactor units, as well as advanced materials development.

Other NE programs conduct R&D that supports the nuclear reactor R&D efforts. The Fuel Cycle R&D program—which is primarily responsible for developing sustainable nuclear fuel cycles,[14] innovative processes to recover uranium from spent nuclear fuel, and nuclear fuel storage options—also supports existing reactors by developing accident-tolerant fuel and advanced reactors by developing advanced proliferation-resistant fuels.[15] The Nuclear Energy Enabling Technologies program conducts R&D on technologies that directly support and complement NE's advanced reactor and fuel cycle R&D efforts. This program coordinates R&D on common issues and challenges that confront other NE R&D efforts to avoid duplication of effort.[16]

NE conducts most of this research at 10 DOE national laboratories across the country and, over the past 3 years, has spent an average of about $840 million per year on its mission to advance nuclear power (see table 1).[17] Of this amount, NE provides about $50 million annually to engage U.S. universities through its Nuclear Energy University Program to fund R&D and to build infrastructure and capabilities to enhance universities' ability to perform research and educate students.[18]

Table 1. Total Funding for DOE Office of Nuclear Energy (NE), Fiscal Years 2012 through 2014

Dollars in millions			
Program	**Fiscal year 2012**	**Fiscal year 2013**	**Fiscal year 2014**
Fuel Cycle Research and Development	$181.0	$169.9	$186.2
Nuclear Energy Enabling Technologies	71.3	67.9	71.1
Reactor Concepts Research, Development, and Demonstration	110.7	104.8	112.8
Small Modular Reactor Licensing Technical Support	67.0	62.7	110.0
Other NE programs, transfer from Department of State, and use of prior year balances[a]	423. 9[b]	393.0[b]	397.4
Total	**$853.8**	**$798.3**	**$877.5**

Source: DOE. | GAO-14-545

Note: Numbers may not add to total due to rounding.

[a] Other NE programs include facilities management, international cooperation, the Idaho Sitewide Safeguards and Security program, the Integrated University Program, and overall direction and execution of the Office of Nuclear Energy programs. The Integrated University Program funding is intended to support nuclear science and engineering by funding scholarships and fellowships in nuclear energy related fields of study, among other things. These funding levels

also include funds transferred from the Department of State, funds transferred to the Small Business Innovation Research and Small Business Technology Transfer programs outside of NE, and the use of NE's prior year balance.

[b] The Idaho Sitewide Safeguards and Security program was funded within Other Defense Activities for fiscal years 2012 and 2013, but are reflected this table. The program was funded within NE for fiscal year 2014. Other Defense Activities funds elements that relate to and support the defense oriented activities within DOE.

NE's recent advanced reactor R&D efforts began in 2000, when NE convened a group of senior governmental officials from nine countries to discuss the development of such reactors in the United States and internationally.[19] This group, called the Generation-IV International Forum, and the Nuclear Energy Advisory Committee produced a Technology Roadmap for Generation-IV Nuclear Energy Systems in 2002. The intent of the forum was to develop competitively priced and reliable nuclear reactors, while satisfactorily addressing nuclear safety, waste, and proliferation concerns. In response to these efforts, the United States determined that it would fund the development of a high-temperature gas-cooled reactor as its top priority.[20] In addition, NE funded research into sodium-cooled fast reactors—fast reactors that use sodium to cool the reactor core—as well as a variety of other advanced reactors.[21]

The Energy Policy Act of 2005 (EPAct 2005) established in law the Next Generation Nuclear Plant (NGNP) Project.[22] The purpose of the NGNP is to develop a prototype reactor using advanced technology to generate electricity, hydrogen, or both. The law states that the NGNP Project shall consist of research, development, design, construction, and operation of a nuclear reactor prototype, and specifies completion dates for the project's two phases, as well as certain other requirements, including licensing. Specifically, EPAct 2005 states that Phase 1 of the project, which requires DOE to, among other things, conduct R&D activities enabling it to select and validate an appropriate technology, culminating in the selection of a technology and initial design parameters, by a target date of September 30, 2011. The law also authorized DOE to submit a report to Congress identifying an alternative date upon which the agency would select the technology and initial design parameters. Phase 2 of the project, in which DOE would develop a final design for a nuclear reactor prototype, apply for licenses to construct and operate the reactor technology, construct the prototype, and begin operations, is to be completed by a target date of September 30, 2021, although DOE is again authorized to submit a report establishing an alternate date for completion. EPAct 2005 also mandated the organization of a consortium of appropriate industrial partners that will carry out cost-shared R&D, as well as design, construction, and operation on behalf of the NGNP Project, and that the NGNP prototype reactor be located at the Idaho National Laboratory in Idaho Falls, Idaho.

In February 2006, the Nuclear Energy Advisory Committee recommended accelerating the NGNP Project schedule to, among other things, make the project more attractive to industry.[23] However, also in February 2006, the administration announced the newly formed Global Nuclear Energy Partnership program, which sought to encourage the expansion of nuclear energy while addressing the burden of spent fuel disposal and the risk of nuclear weapons proliferation and, according to a 2008 National Academy of Sciences report, led to reduced funding for the NGNP Project. Under the Global Nuclear Energy Partnership program, DOE focused advanced reactor R&D activities on developing sodium-cooled fast reactors, and it changed its approach from designing and building a small engineering-scale demonstration of the reactor and reprocessing facility, led by DOE's national laboratories, to accelerating its work with industry to demonstrate commercially viable sodium-cooled fast reactor technology in full-scale facilities.[24]

In 2008, we reviewed NE's Global Nuclear Energy Partnership program and found that DOE's original approach to the domestic component of the program—building engineering-scale facilities—would meet the program's objectives if the advanced spent nuclear fuel recycling technologies on which it focused could be successfully developed and commercialized.[25] However, we also reported that the approach lacked industry participation—potentially reducing the prospects for eventual commercialization of the technologies. NE favored an accelerated approach of building full-scale facilities that would likely require using unproven evolutions of existing technologies that would reduce the long-term benefits of the sodium-cooled fast reactor.[26] We recommended that NE reassess its preference for an accelerated approach. In response, DOE decided, in 2009, to no longer pursue the Global Nuclear Energy Partnership program. However, NE continued research related to sodium-cooled fast reactors, with a new focus on long-term R&D coordinated with NE's fuel cycle research.

Also in 2008, the National Academy of Sciences' National Research Council issued a report reviewing NE's R&D efforts and concluded that the success of any particular advanced reactor technology in the United States would depend on policy decisions and other factors beyond NE's control.[27] In addition, the report concluded that NE's resources were barely adequate for basic studies related to the NGNP Project and entirely inadequate for (1) exploring the sodium-cooled fast reactor at a research level and (2) investigating other reactor technologies. The report also stated that selecting a specific technology to develop from among the options known at that time would have been premature.

In April 2010, NE issued its *Nuclear Energy Research and Development Roadmap: A Report to Congress*, which provides a basis to guide NE's internal programmatic and strategic planning for research going forward. Also in April 2010, NE issued *Next Generation Nuclear Plant: A Report to Congress*, which presents the historical background of the project; details the project's spending; and discusses the principal investments in design, licensing, and research. As discussed in this report, NE selected the high-temperature gas-cooled reactor as the advanced reactor technology to develop under the NGNP Project.

In its 2011 Phase 1 review of the project, the Nuclear Energy Advisory Committee reported that the NGNP Project was not ready to proceed to the complete set of Phase 2 activities, citing

- the need for more detailed design and R&D;
- the need to resolve key licensing issues;
- constraints imposed by EPAct 2005, such as the Idaho site requirement;
- the absence of a public-private partnership, as required;
- absence of industrial partners willing to commit to share in the cost of constructing the prototype reactor; and
- an unrealistic project plan given the limited amount of conceptual design work completed.

The committee recommended that NE continue conducting its Phase 1 R&D, focused on one technology and completing design work, initiate the partnership with the nuclear industry, and continue to engage the Nuclear Regulatory Commission (NRC)—the federal agency that licenses and regulates the nation's civilian use of nuclear reactors—to ensure that the regulatory framework for this new reactor technology would be ready to support commercialization. The committee also recommended eliminating the requirement that the NGNP prototype be sited at Idaho National Laboratory.

In October 2011, DOE submitted a letter to Congress in response to EPAct 2005 requirements to transmit the committee's report and to report on certain requirements to complete Phase 1 of the NGNP Project. As noted above, EPAct 2005 required, by September 30, 2011, that DOE select technology and initial design parameters; alternately, DOE was to submit a report to Congress identifying a new date upon which the agency would do so. In the 2011 letter, DOE stated that it was notifying Congress that the department had not selected the initial design parameters for the NGNP by September 30, 2011, and that it would not proceed with Phase 2 design activities at that time. The letter also stated that DOE would focus on remaining applied R&D, work with NRC on licensing framework, and establish the public-private partnership—in essence, to follow most of the advisory committee's recommendations—until conditions favorable to completing the NGNP Project warranted a change in direction. Further, DOE asserted that the partnership—rather than DOE—would select initial design parameters and would provide an update to the project's schedule and milestones.

In its fiscal year 2014 budget request to Congress, DOE indicated that NE would continue to fund some NGNP research activities under the Advanced Reactor Concepts subprogram.

NE's Approach to Advanced Reactor R&D Supports Several Advanced Reactor Technologies

NE's approach to its advanced reactor R&D is to support research on technologies associated with three main types of advanced reactors: high-temperature gas-cooled reactors; liquid-metal-cooled fast reactors, including the sodium-cooled fast reactor; and fluoride-salt-cooled high-temperature reactors. NE also conducts research supporting other less-developed advanced reactor technologies and supports the development of advanced small modular reactors. NE's approach to advanced reactor R&D addresses broad programmatic goals—including improvements in the economics, safety, and proliferation resistance of nuclear power plants—and aims to develop technologies to reduce nuclear waste and greenhouse gas emissions. This approach provides several advantages, including flexibility in responding to changes in future U.S. energy policy or other circumstances.

NE's Advanced Reactor R&D Efforts Are Focused on Three Primary Reactor Technologies

NE's approach to its advanced reactor R&D efforts primarily is to support research on three main advanced reactor technologies: high-temperature gas-cooled reactors; liquid-metal cooled fast reactors, including the sodium-cooled fast reactor; and fluoride-salt-cooled high-temperature reactors. On a smaller scale, NE also conducts or funds research supporting other less-developed advanced reactor technologies. In addition, NE conducts research supporting a variety of technologies related to the development of advanced small modular reactors. In discussions with representatives from the nuclear industry, members of the National Academy of Sciences' National Research Council, and others, we found that views frequently varied on which specific technologies NE should be supporting through its R&D efforts.

High-Temperature Gas-Cooled Reactors

High-temperature gas-cooled reactors produce energy in the form of high-temperature heat—which can produce electricity or be used as process heat—and differ from existing light water reactors in three key features: using (1) helium gas instead of water as a coolant; (2) graphite instead of water to slow neutrons and sustain the nuclear reaction; and (3) advanced nuclear fuel, which offers safety benefits at high temperatures. These three features make high-temperature gas-cooled reactors capable of operating at higher temperatures than existing light water reactors, thus offering a broader range of applications to industrial processes, as well as higher heat-to-electricity energy conversion efficiencies than are achievable with the lower operating temperatures of light water reactors. The high-temperature gas-cooled reactor is an advanced reactor technology that is expected to be helpful in limiting greenhouse gas emissions, according to NE officials, DOE laboratory staff, and NE documents. In addition, the technology offers inherent and passive safety features—including advanced fuel, helium coolant, and passive heat removal—that are especially important in a post-Fukushima world, according to NE officials.

NE's current involvement with high-temperature gas-cooled reactor R&D began in 2002, when, under the Generation-IV International Forum, high-temperature gas-cooled reactor systems were selected as one of six advanced reactor technologies to be developed by the international consortium. The United States was one of six countries, plus the European Union, that took the lead in developing this reactor technology, according to NE officials. NE chose to pursue high-temperature gas-cooled reactors because, according to NE officials, laboratory staff, and NE documents, they met the criteria of its advanced reactor programs, including potential improvements over existing light water reactors in safety, economic viability, and their promise for reducing greenhouse gas emissions. NE officials told us that the high-temperature gas-cooled reactor was also chosen because it had operating history in the United States and throughout the world, which meant that many important technical details had already been resolved. At the time, NE concluded that its research would build on existing operational experience, thus increasing the likelihood that high-temperature gas-cooled reactors could be developed and commercialized. NE envisioned its research would further demonstrate the technical and economic viability of the high-temperature gas-cooled reactor technology.

Pursuant to EPAct 2005, NE is to undertake research, development, design, construction, and operation of an advanced nuclear reactor prototype with a targeted completion date of September 30, 2021.[28] From 2005 to 2011, under Phase 1 of the NGNP Project, NE spent more than $500 million on R&D in support of high-temperature gas-cooled reactor technologies. This research was conducted primarily at DOE's national laboratories, and it focused on preliminary reactor design work, developing new reactor fuel, and testing high-temperature materials. The NGNP Project also collaborates with universities and industry on some R&D activities, and NE has coordinated with NRC to conduct R&D necessary to design and license high-temperature gas-cooled reactors in the United States. The last high-temperature gas-cooled reactor was licensed in the United States in 1973, and many questions remain about the process by which NRC will consider license approval for these reactors. According to NE and NRC officials, NE began consulting with NRC in the mid-2000s as NRC worked to (1) update its policy on the regulation of advanced reactors and (2) produce a report to Congress on its advanced reactor licensing strategy for NGNP, which was jointly published in 2008. During this time, NE shared detailed technical information and supporting materials with NRC, information meant to mitigate risks associated with the licensing process, according to NE and NRC officials. Although DOE determined that it would not proceed to Phase 2 design activities for the NGNP Project in 2011, NE continues to fund research on some aspects of a high-temperature gas-cooled reactor, including the testing of advanced reactor fuel and high-temperature materials including graphite.

We found varying views on the potential economic feasibility of the high-temperature gas-cooled reactors. Some industry representatives and a member of the National Academy of Sciences' National Research Council that we interviewed questioned the appropriateness of NE continuing to fund high-temperature gas-cooled reactor research, citing concerns over the reactor's economic viability in relation to the current cost of natural gas. One industry representative cited studies showing that the technology is not economically viable, as well as the fact that NE has, to date, been unable to get industrial partners interested in sharing project development costs. However, an economic analysis by the NGNP Industry Alliance—an international consortium of potential end users, owner-operators, and technology companies brought together to partner with NE and commercialize the high-temperature gas-cooled reactor—concluded that the high-temperature gas-cooled reactor would be economically viable under certain market conditions. In addition, NE has conducted a series of feasibility studies, including detailed economic analyses and studies of potential industrial applications for the process heat, to demonstrate conditions under which the high-temperature gas-cooled reactor technology becomes economically competitive. Representatives from the NGNP Industry Alliance said they believe that, with favorable trends in natural gas prices, the high-temperature gas-cooled reactor will be economically viable by the time the prototype reactor is built in the early-2030s, as is projected under the NGNP Industry Alliance's current project time frames.

Liquid-Metal-Cooled Fast Reactors

The liquid-metal-cooled fast reactors, including sodium-cooled fast reactors, use liquid-metal to cool the reactor core. In addition to producing electricity, a primary benefit of these reactors is in nuclear waste management. Fast reactors can use reprocessed spent nuclear fuel as their energy source, which would help the United States reduce the amount of spent fuel from light water reactors that would need to be stored or eventually placed in a geologic repository. Fast reactors also minimize nuclear waste generation by significantly improving fuel use efficiency as compared to traditional light water reactors.

NE identified the sodium-cooled fast reactor as a reactor technology of interest in 2002 under the Generation-IV International Forum because, according to NE officials, it met criteria of its advanced reactor programs, including significant advances in proliferation resistance, its potential for improving the sustainability of the nuclear fuel cycle, and its management of highly radioactive waste elements. NE continues to conduct research in support of sodium-cooled fast reactors because of their advantages for addressing nuclear waste and because reactors with the same basic technology have been built and operated in the United States and around the world since the 1960s. In fact, several other countries, including Japan and France, had or currently have operating sodium-cooled fast reactors.

We also found varying views on sodium-cooled fast reactors among industry representatives, members of the National Academy of Sciences' National Research Council, and others that we interviewed. Some members of the National Academy of Sciences' National Research Council and an industry representative cited concerns over the safety of sodium-cooled fast reactors—including the highly reactive nature of the sodium in the presence of water and the threat of sodium fires—and believe that these safety issues may never be fully overcome. Moreover, some industry representatives and members of the National Academy of Sciences' National Research Council told us that a fast reactor technology market does not exist in the United States, citing other more cost-effective options for storing spent nuclear fuel, including storage in aboveground casks or water pools as is the current practice at U.S. nuclear power reactors. These individuals believe that NE's sodium-cooled fast reactor research is ill-advised.

In contrast, NE officials and some industry representatives that we interviewed believe that remaining technical challenges with the sodium-cooled fast reactor can be overcome. In addition, NE officials said it is conceivable that changes in government policy for handling spent nuclear fuel in the United States will create a market for fast reactors, as it has in the United Kingdom. Moreover, NE believes that the sodium-cooled fast reactor, or other similar fast reactor technology, may be instrumental in efforts to develop a sustainable nuclear fuel cycle. In 2011, NE undertook a comprehensive study of various options for improving the sustainability of the nuclear fuel cycle, which would potentially entail using fast reactor technology and waste reprocessing to create nuclear power systems that better manage and reduce the generation of nuclear waste when compared to a once-through light water reactor fuel cycle. Although this study did not consider specific advanced reactor technologies, a closed fuel cycle may require using an advanced fast reactor technology, such as the sodium-cooled fast reactor.

Fluoride-Salt-Cooled High-Temperature Reactor

The fluoride-salt-cooled high-temperature reactor design takes advantage of the physical characteristics of liquid-salt coolant to enable the development of a high-temperature system that is scalable to larger power and able to operate at lower pressure and higher power density than the helium-cooled high-temperature gas-cooled reactor. With these characteristics, according to NE officials and DOE documents, these reactors could provide potential safety benefits and increased efficiency over existing light water reactors while maintaining the benefit of providing both electricity and process heat for industrial applications. NE officials told us that this mix of characteristics is the reason why NE provides limited funding research into fluoride-salt-cooled high-temperature reactor. However, the fluoride-salt-cooled high-temperature reactor technology is not very mature or well tested, and because of this is considered more of a long-range advanced reactor technology, according to NE officials. NE funds R&D into fluoride-salt-cooled high-temperature reactors mainly through NE's Nuclear Energy University Program, with the research mainly conducted at universities across the country. Two industry representatives we interviewed took issue with NE for funding a reactor technology that is unproven and that in their view has little chance of ever being built. In response, NE officials told us that the potential benefits of fluoride-salt-cooled high-temperature reactor over other advanced reactor technologies warrant providing limited funds and utilizing university research capabilities.

Other Advanced Nuclear Reactor Technologies

NE conducts or funds R&D on other advanced reactor technologies on a small scale, mainly to assess their potential and better characterize their performance capabilities. NE officials told us that most of these technologies have some unproven aspects, operate in novel ways, or have other characteristics that increase the risk associated with their development. NE supports research into some of these potentially transformative, long-term technology options through its Nuclear Energy University Program. For example, in 2013, the program funded research to assess the feasibility of an advanced reactor fueled with depleted uranium, a design offering a 30-fold increase in uranium ore utilization verses contemporary light water reactor designs.

NE also funds research into promising advanced reactor technologies through the Advanced Reactor Concepts Technical Review Panel process. Through this process, NE identifies R&D needs for potentially viable advanced reactor technologies to inform NE advanced reactor R&D funding decisions. A goal of the process is to facilitate greater engagement between DOE and the nuclear industry. NE first solicited information on advanced reactor proposals from industry in February 2012, after which a review panel made up of experts from national laboratories, universities, and industry reviewed the proposals against established evaluation criteria, including safety, market attractiveness, economics, proliferation risk, waste generation, security, and potential regulatory challenges. The panel's assessment of market attractiveness focused on the proposed technologies' ability to be competitive in the marketplace, and it included variables like efficiency, initial capital costs, and economic factors such as construction, manufacturing, and operating costs and uncertainties, as well as the resulting cost of electricity, according to the Technical Review Panel report. The objective of the Technical Review Panel process was to evaluate the viability of the technologies, understand the R&D needs of each, and prioritize research to support development and commercialization of each. After R&D needs and priorities were identified, NE issued a funding opportunity announcement, competitively selected four projects, and provided a total of $3.5 million in funding for those projects, according to NE officials. Many nuclear industry representatives we interviewed applauded NE's effort and told us that this process was an effective way for NE to collaborate with industry and that it begins to address a long-standing industry concern that NE's R&D efforts did not coordinate with industry or meet industry needs. However, these industry representatives also stated that the $3.5 million in R&D funding was insufficient to meaningfully address the need for collaboration between NE and industry as it was enough to fund a very small number of R&D activities. Notably, Congress provided NE with an additional $12 million to support a continuation of this effort. According to NE officials, NE has issued another industry solicitation and will use the information gathered to make additional industry cost-shared R&D awards early in fiscal year 2015.

Advanced Small Modular Reactors

NE also conducts research on advanced small modular reactors with the goal of supporting the development of innovative small modular reactor designs that offer improved safety, functionality, and affordability. These R&D efforts support advanced small modular reactors that offer simplified operation and maintenance, more efficient energy conversion, and increased proliferation resistance and security. More specifically, NE funds research on advanced sensors, instrumentation and controls, control systems for multiple units, advanced materials, and other major system components. In addition, NE funds efforts to create standards and codes for small modular reactor materials to support the eventual licensing of these advanced reactor technologies.

In 2012 and 2013, through its Small Modular Reactor Licensing Technical Support program, NE issued funding opportunity announcements for cost-sharing with industry for the development of small modular reactor designs—including small modular reactors based on light water reactor technology, as well as advanced small modular reactors—to support the program's vision to provide additional nuclear power options that offer more flexibility in financing, siting, and end-use applications than large light water reactor designs. Under the cost-sharing arrangement for each funding opportunity, DOE is supporting design development, first-of-a-kind engineering, experiments, and analysis in support of gaining design certification approval from NRC for the small modular reactors so that commercial deployment of the first small modular reactor can begin. Industry proposals under these announcements were judged by independent selection panels based on a series of criteria, including the extent to which the design incorporates safety, operability, efficiency, economics, and security characteristics that exceed the capabilities of current reactor designs; the likelihood of expeditiously achieving design certification and deployment;[29] the overall quality of the project plan and business approach; and other factors. Based on our review of the two funding opportunity announcements, however, we found that the funding opportunity announcements differed in the economic information that NE required for proposals. In the 2012 announcement, NE more directly addressed economics and marketability by requiring applicants to propose business plans "to meet expanding domestic electricity requirements at a competitive price" and to "provide their plan to achieve successful commercial deployment" of the technology. By contrast, the 2013 announcement indicated that the economic criteria used to evaluate proposals would be based on the designs' construction, fabrication, deployment, and operational costs. NE officials told us that these criteria indirectly assess economics and marketability of these technologies, and that the type of economic information received from applicants in 2012 was very preliminary and did not provide a good discriminator with which to evaluate proposals. In addition, these officials stated that it was incumbent on the applicants to ultimately assure marketability as they were providing most of the funding and have a profit motive. For both funding opportunity announcements, the panel's evaluation resulted in choosing a small modular reactor design based on conventional light water reactor technology. Some industry representatives, members of the National Academy of Sciences' National Research Council and the Nuclear Energy Advisory Committee, and NE officials told us that this selection was an appropriate choice because it has a significantly better chance of being licensed and constructed in the required time frame, as compared to advanced small modular reactor designs that are not based on conventional light water reactor technology.

NE's Approach Addresses Broad Programmatic Goals and Policy Objectives, and Provides Flexibility in Responding to Changes in U.S. Energy Policy

While the broad goals of NE's advanced reactor R&D efforts are to improve the economics, safety, and proliferation resistance of nuclear power plants, the R&D efforts also aim to develop advanced reactor technologies that can prepare the United States to address policy objectives such as reducing nuclear waste and greenhouse gas emissions. NE's approach to advanced reactor R&D is to conduct research in support of multiple advanced reactor technologies. According to NE officials and documents, because NE's approach to advanced reactor R&D has multiple goals and seeks to address several different policy objectives, NE works on multiple technologies simultaneously. A key objective of NE's advanced reactor R&D efforts is to conduct research to remove technology barriers or reduce technology risks, while collaborating with industry and academia, with the ultimate goal for industry to take the results of NE's research to the next step of development and commercialization. NE focuses on R&D that industry does not have the means to carry out, according to NE officials, with the expectation that the research will reduce financial risks to industry and thereby increase the affordability of industry investment in new nuclear technologies. In addition, NE engages and collaborates with NRC on issues related to the eventual licensing of advanced reactors, including understanding the likely scope and extent of R&D necessary to support the licensing process.

While advanced reactors are attractive for many reasons, NE carries out research on a variety of reactors because, in part, different reactor types can address particular objectives, according to NE officials. For instance, fast reactors can be better at addressing the nuclear waste issue than some other advanced reactors, while high-temperature gas-cooled reactors provide process heat and may be a better solution for addressing greenhouse gas emissions. According to NE officials, the development of fast reactors, such as the sodium-cooled fast reactor, is likely to play a critical role in managing spent nuclear fuel if and when the United States decides to reprocess and use its spent nuclear fuel rather than store it at reactor sites,[30] as is the current practice at U.S. nuclear power reactors, or isolate it in a geologic repository underground, as has been proposed.[31]

To remain aware of industry's R&D needs and international nuclear energy developments, NE regularly collaborates with industry and international organizations, according to NE officials and NE documents. NE officials told us that NE regularly collaborates with industry on specific R&D projects by sharing technical data and information. For example, NE is currently collaborating with industry on advanced fuels and materials, among other things. NE officials told us they work with industry and have conversations regarding specific R&D activities. According to NE officials and some industry representatives that we interviewed, this type of collaboration has been increasing in recent years, and one industry representative stated that such collaboration is critically important to ensuring that NE's activities are relevant for industry.

One way that NE has recently collaborated with industry, according to NE officials, was through the Advanced Reactor Concepts Technical Review Panel process. Some industry representatives we talked to stated that this review panel process was beneficial to both industry and NE because it helped inform NE of industry R&D needs and because it has provided some funds to industry to carry out research on promising new technologies. Industry officials also told us that the process has opened up some new channels of communication between NE and industry. However, industry representatives also stated that, although this collaboration with industry is beneficial, NE could be doing more to ensure that its R&D is more fully aligned with industry needs. For example, according to one industry official, NE conducts some research that, while interesting and potentially beneficial, has little utility for industry's current needs.

NE carries out international collaboration through ongoing meetings of the Generation-IV International Forum and through the International Atomic Energy Agency,[32] through the Organisation for Economic Co-operation and Development (OECD),[33] and through bi-lateral agreements with many countries around the world, including Canada, the Russian Federation, the

People's Republic of China, Japan, the Republic of Korea, and countries in the European Union. NE officials cited several examples of such collaboration, such as with the People's Republic of China on high-temperature gas-cooled reactors; with France on their sodium-cooled fast reactor development project; with Japan on advanced materials for sodium-cooled fast reactors; and with the Republic of Korea on sodium-cooled fast reactors.

NE's approach to advanced reactor R&D provides several advantages, primarily flexibility in responding to changes in future U.S. energy policy or other circumstances, according to NE officials. The officials said they believe that conducting research in support of multiple advanced reactor technologies gives the agency the flexibility to respond to external factors affecting the direction of their advanced reactor R&D efforts, including changes in U.S. energy policy, energy markets, or other areas. Specifically, NE officials told us that the current approach positions NE to respond to changes in U.S. energy policies, such as policies for managing the nation's nuclear waste or controlling greenhouse gas emissions. Changes in either of these policies would affect the direction of NE's advanced reactor efforts, according to NE officials. For instance, a policy calling for the United States to manage nuclear waste by reprocessing spent nuclear fuel and reusing it as reactor fuel would result in NE focusing efforts and concentrating resources on developing and deploying fast reactor technologies. NE officials told us that the current practice of storing nuclear waste in aboveground facilities at nuclear power plants across the country will eventually be changed, and waste will either be moved to long-term underground repositories, or reprocessed and burned in fast nuclear reactors.[34] NE officials told us that ongoing research into fast reactor technologies is important so that NE is positioned to react to changes in U.S. policy toward the handling of nuclear waste, including waste that has already been generated and waste that continues to be generated.

Similarly, NE officials said that policies that affect the prices of various energy sources would have an effect on the commercial attractiveness of high-temperature reactors, including high-temperature gas-cooled reactors. For instance, officials cited the possibility of the imposition of a carbon tax to control greenhouse gas emissions, or the possibility of natural gas prices rising, either of which would make nuclear energy more economically competitive and increase the attractiveness of the high-temperature gas-cooled reactors that produce both electricity and process heat for industrial applications. These industrial applications currently rely heavily on natural gas or coal plants as their source for high-temperature process heat. NE officials stated that natural gas prices in other countries are already at levels where the high-temperature gas-cooled reactors are projected to be economically competitive and that this has resulted in some interest from outside the United States in the development of this technology

According to NE officials, another advantage of NE's approach to advanced reactor R&D is that NE is able to maintain an employee base with knowledge and expertise on a wide variety of reactor technologies. NE officials told us that maintaining staff expertise is important so NE can continue to conduct research on the various technologies, train the next generation of scientists and engineers, and be ready to support the production of prototype reactors when the time comes. In addition, maintaining a level of expertise in a variety of advanced reactor technologies means that NE can engage with, monitor, and support other countries as they develop advanced reactor technologies. These officials said this is important because other countries are actively developing advanced reactor technologies, and the United States needs scientists that can understand how those reactors operate, in part, to judge their safety and nuclear proliferation risks.

NE officials also said that conducing R&D on several types of advanced reactors simultaneously, rather than focusing on a single reactor type, also gives NE the ability to fund R&D supporting promising but unproven reactor technologies. For instance, NE is funding limited research on lead-cooled fast reactors, which offer the potential for improved safety and proliferation resistance over other advanced reactor technologies, but they have some unproven technologies and components, according to NE officials. Similarly, NE is funding research in support of an advanced small modular reactor based on fast reactor technology that would potentially address the nuclear waste issue and also provide process heat for industrial applications. NE officials said that it is important to have funds available to support these and other potentially game-changing technological breakthroughs. However, in its June 2013 report, the Nuclear Energy Advisory Committee was critical of NE's approach, saying that NE needs to better prioritize its R&D efforts on a smaller number of advanced reactor technologies to focus research funding on the ultimate goal of deploying an advanced reactor prototype.

Although NE selected the technology to develop under the NGNP Project, many members of the National Academy of Sciences' National Research Council, members of the Nuclear Energy Advisory Committee, and industry representatives we interviewed agree with NE's approach to advanced reactor R&D because the time is not right for NE to move to the deployment phase. For instance, representatives from industry and the Nuclear Energy Advisory Committee told us that uncertainties around current policies for handing nuclear waste and controlling greenhouse gases do not make a compelling case for choosing an advanced reactor technology to deploy as a prototype. The 2008 National Academy of Sciences' National Research Council review of NE's advanced reactor R&D efforts agreed with NE's approach to advanced reactor R&D, saying that there are several policy matters and other questions—undetermined nuclear waste management options, unformulated environmental policy, ongoing work of other countries on advanced technologies, and unclear nonproliferation regimes—that will affect NE's decisions and priorities. This review team stated that, given these unknowns, it would be premature to select a winning technology from among current options.

In addition, in January 2012, the President's Blue Ribbon Commission on America's Nuclear Future recommended having the United States continue multiple near-term (i.e., light water reactor) and long-term (e.g., small modular reactor, sodium-cooled fast reactor, high-temperature gas-cooled reactor) R&D efforts until NE could defensibly select technologies that would meet certain regulatory and policy requirements (e.g., safety, environmental protection, security, and nonproliferation). Moreover, members of the Nuclear Energy Advisory Committee and the National Academy of Sciences' National Research Council, and representatives from industry told us that current NE funding levels would prohibit NE from deploying a prototype reactor even if NE chose an advanced technology to deploy. Some of them said that NE is correctly positioning itself to be prepared to deploy a prototype reactor in the long-term as policies or energy markets change. One Nuclear Energy Advisory Committee representative said that United States could focus its advanced reactor R&D efforts quickly in response to a policy change or other congressional direction, provided that NE also saw increased funding.

NE Uses Internal and External Reviews to Set Program and Funding Priorities but Does Not Have a Strategy for Deploying an Advanced Reactor Prototype

NE's uses internal and external reviews to set program and funding priorities for advanced reactor R&D and to evaluate progress toward program goals. However, NE does not have a strategy for overcoming barriers to deploying an advanced nuclear reactor prototype, increasing the likelihood that such a reactor will not be built by the 2021 target date specified in EPAct 2005. Not deploying a prototype carries certain risks, including waning U.S. influence in the safe operation of nuclear plants internationally and potential loss of certain knowledge and expertise.

NE Plans, Prioritizes, and Evaluates Its R&D Activities through Internal and External Reviews

NE takes a number of steps to plan and prioritize its advanced reactor R&D efforts and evaluate progress toward program goals. Before its annual program planning meetings, NE and national laboratory staff develop a list of R&D efforts considered to be priorities. NE management reviews this information in light of program goals, including long-term goals described in NE's 2010 R&D Roadmap, program funding, and schedules, according to NE officials. Once the research priorities are established and approved by management, the individual laboratories develop detailed work plans, which describe the objectives and scope of the work to be performed. These work plans are reviewed to ensure that the proposed work is aligned with NE's mission and that the work can be accomplished within the allotted budget and time frames, according to NE officials. All of the approved work plans are then entered into NE's performance management system—the Program Information Collection System—which allows NE to track progress toward budget and schedule milestones on an ongoing basis. According to a laboratory staff member, this system tracks progress toward short-term goals—on a monthly basis—and long-term goals—on yearly, 3-year, and 5-year time frames.

NE monitors and evaluates its advanced reactor R&D activities on an ongoing basis through the Program Information Collection System and conducts program reviews—monthly, quarterly, and annually—to assess progress toward program goals, according to NE officials. For example, officials from the Advanced Reactor Concepts and Advanced Small Modular Reactor subprograms hold monthly progress review meetings to discuss, among other things, program updates, technical highlights, and budget and milestone status updates. The officials use monthly status tracking reports generated by the performance management system as part of these reviews, in which officials review cost and schedule performance data. In addition to the monthly meetings, officials from the Advanced Reactor Concepts and Advanced Small Modular Reactor subprograms typically conduct four in-depth reviews of each year, according to NE officials. These reviews focus on one or more specific areas of research, and officials discuss progress toward goals, important issues or problems, and plans going forward. For example, the meeting minutes from the quarterly review of the Advanced Reactor Concepts and Advanced Small Modular Reactor subprograms in July 2013 show that officials discussed accomplishments and also priorities for the upcoming fiscal year, and conducted in-depth discussions of certain program areas and overviews of others. In addition, NE conducts annual reviews of activities across multiple subprograms and topics to ensure that NE's efforts are complementary and nonduplicative, and also to gain insight into areas of potential collaboration. For example, during its annual review of the nuclear reactor R&D efforts in March 2014, NE officials discussed progress on fuels for the high-temperature gas-cooled reactor, advanced reactor licensing, and advanced reactor materials for small modular reactors, among other things. On a less-formal basis, management officials at the national laboratories are in frequent communication with NE management through weekly teleconferences to provide regular progress updates and provide information on unforeseen circumstances or challenges, according to NE officials.

NE also takes steps to coordinate efforts across its R&D programs and subprograms to leverage experience and funding, as well as to reduce redundant R&D activities. Officials from the Advanced Reactor Concepts and Fuel Cycle subprograms stated that they frequently coordinate with each other because their R&D efforts are interdependent. For example, the Fuel Cycle subprogram is conducting R&D on accident tolerant fuels that will be used for advanced reactors, so coordination between the Fuel Cycle subprogram and the Advanced Reactor Concepts subprogram is crucial, according to NE officials.

To further help ensure that R&D efforts are coordinated and to minimize redundancies, NE established the Nuclear Energy Enabling Technologies program in fiscal year 2011. The program is designed to conduct R&D on crosscutting technologies that complement NE's activities to support and enable the development of new advanced reactor designs and fuel cycle technologies. NE created the program to better coordinate and integrate R&D activities after officials identified some similar efforts being performed on crosscutting areas, such as materials, across more than one program, according to NE officials. Through this program, NE has awarded over $9 million to support R&D projects focused on reactor materials, advanced sensors and instrumentation, and advanced methods for manufacturing, among other things. NE determines which R&D efforts are conducted by the Nuclear

Energy Enabling Technologies program by reviewing R&D proposals submitted by different groups, including the national labs, universities, research institutions, and industry, according to NE documents. Specific R&D projects are selected based on common needs of programs and subprograms, with each selected project required to support at least three programs or subprograms.

Others also periodically conduct external reviews of NE's advanced reactor R&D to inform the planning and prioritization efforts for and assess the progress of its R&D activities. Most prominently, according to NE officials, the Nuclear Energy Advisory Committee provides NE with independent advice and recommendations on complex science and technical issues that arise in planning, managing, and implementing NE's R&D activities. The Nuclear Energy Advisory Committee typically meets twice annually with NE management to discuss its reports and recommendations. The Nuclear Energy Advisory Committee's subcommittee on Nuclear Reactor Technology is intended to provide expert guidance to NE on both the short-term and long-term direction of its R&D efforts on reactor technologies. NE officials and Nuclear Energy Advisory Committee representatives told us that the committee has been beneficial in providing expertise to NE and that NE has been responsive to the committee's recommendations. In 2011, a Nuclear Energy Advisory Committee review of NE's R&D efforts on the NGNP Project determined that NE should not move forward with the complete set of Phase 2 activities of the project, citing constraints imposed by EPAct 2005 and difficulties finding industry partners. Subsequently, in 2011, NE informed Congress that it would not proceed with Phase 2 design activities of the NGNP Project until circumstances warranted a change in direction.

NE's efforts have also been reviewed by other outside entities, including the Secretary of Energy's Advisory Board, which provides advice and recommendations to the Secretary of Energy on various topics. For example, in 2012, the Secretary of Energy requested the board identify areas in which standards for safety, security, and nonproliferation should be developed for small modular reactors; identify challenges, uncertainties, and risks to commercialization; and provide advice on approaches to manage these risks and accelerate deployment of these reactors. The board determined that the commercialization of small modular reactors was likely to produce multiple benefits for the country, including helping provide for a more reliable power grid with more widely distributed power generation once current light water reactors are retired; supporting clean generation and reduced carbon emissions; and helping preserve influence of the United States on nuclear nonproliferation issues. The board stated that to deploy small modular reactors widely, the nation must develop a robust small modular reactor industry that can manufacture cost-competitive small modular reactors that meet U.S. regulatory standards, and that the primary risk for commercialization of these reactors, beyond design certification and licensing, is the ability to drive the plant costs down sufficiently to become competitive with other energy sources, such as natural gas, without compromising safety and security. To develop this industry, according to the board, the U.S. government will likely have to play a significant financial role beyond the Small Modular Reactor Licensing Technical Support program.

NE Does Not Have a Strategy to Overcome Several Barriers to Deploying an Advanced Nuclear Reactor Prototype

Although NE's primary mission is to advance nuclear power through research, development, and demonstration, its deployment of an advanced reactor prototype under the NGNP Project is unlikely in the foreseeable future. Among the different advanced reactor technologies currently supported by NE R&D, the high-temperature gas-cooled reactor technology is the most likely to be deployed and commercialized in the near term, according to an NE planning document. NE officials said that the likelihood is based on the wide range of potential market applications to industry of electricity and process heat and is supported by substantial government investments in the technology's development, including testing of materials, fuels, and other components. NE has consulted with the NGNP Industry Alliance on the project, including discussing the alliance's plan for proceeding with development of a prototype reactor. In addition, NE has done market research on potential industrial applications for the process heat. Further, NE established a contract in 2013 with NGNP Industry Alliance to develop economic analyses detailing how industry may best engage in developing and commercializing high-temperature gas-cooled reactor technologies. According to laboratory staff, development and testing of the advanced fuel for high-temperature gas-cooled reactor has progressed positively, and other research on high-temperature materials and other components has produced positive results.[35]

In 2011, DOE informed Congress that it would not proceed with Phase 2 design activities for the NGNP Project until circumstances warranted a change in direction. According to NE officials, laboratory staff, and representatives of the NGNP Industry Alliance, the NGNP Project remains hindered by several barriers. Specifically, barriers are as follows:

- **Cost-share requirements.** DOE's attempts to implement the cost-share provisions in EPAct 2005 for the NGNP Projects have met with resistance from industry, according to DOE officials and industry representatives, because of differences in how EPAct 2005 is interpreted by NE and by the NGNP Industry Alliance. EPAct 2005 provides that activities by industry must be cost-shared in accordance with the research, development, demonstration, and commercial application cost-sharing provisions established under section 988 of the act. Specifically, the Secretary must require cost-sharing in accordance with this cost-sharing provision when carrying out a research, development, demonstration, or commercial application program or activity that is initiated after August 8, 2005. For applied research and development activities, industry generally is to provide not less than 20 percent of the cost, but the cost-share may be reduced or eliminated if the Secretary determines doing so is necessary and appropriate. For demonstration and commercial application activities, industry generally is to provide not less than 50 percent of the cost, but the cost-share may be reduced if the Secretary determines that doing so is necessary and appropriate considering any technological risk relating to the activity. However, according to NE officials and representatives from the NGNP Industry Alliance, they have been unable to come to an agreement on implementing the cost-share requirement for funding the remainder of the NGNP Project because of disagreement on the applicable cost-share levels and how and when the cost-share would be applied to specific activities or project phases. The NGNP Industry Alliance favors meeting the total cost-share requirement by measuring costs over the course of the remainder of the NGNP Project rather than on an annual basis. According to the NGNP Industry Alliance cost-share proposal from November 2009, the alliance suggested assessing a lower industry cost-share in the first years of the project and increasing the industry share over time, with industry paying the vast share of the annual project costs by the final years of the project. Under this proposed scenario, the alliance states that the cumulative industry contribution would meet the overall cost-share requirement, and NE's portion of the development costs would largely be paid up front. DOE did not fully consider the alliance's proposal for a multiyear approach to the cost-share requirement, according to NE officials, because the project was not proceeding at the time due to funding constraints, competing program priorities, and other factors

Representatives from the NGNP Industry Alliance told us that cost-sharing the development activities on a annual basis is not feasible because it would mean a substantial layout of funds with a very long payoff time and would expose the industry partners to significant financial risks. NGNP Industry Alliance representatives said these risks include unknowns associated with obtaining regulatory approval from NRC for the prototype reactor, and the risk that NE will not be provided sufficient funds through congressional appropriations to meet its obligations. NE officials told us that they understand the NGNP Industry Alliance's perspective and had been attempting to work out an agreement when DOE decided not to proceed to Phase 2 of the project. Representatives from the NGNP Industry Alliance told us that the impasse over cost-sharing needs to be resolved in order to proceed with the NGNP advanced reactor prototype.

- **Site requirement.** According to NE officials, laboratory staff, and NGNP Industry Alliance representatives, the EPAct 2005 requirement that the NGNP reactor prototype be located at the Idaho National Laboratory is another barrier to proceeding with the project. Representatives of the NGNP Industry Alliance said that part of the economic benefit of the reactor prototype would be the use of the high-temperature process heat that results from operating the high-temperature gas-cooled reactor. Alliance representatives said that building the reactor at Idaho National Laboratory foregoes the economic benefit because industries that could potentially use process heat are not located near the laboratory, making the overall prototype reactor less economically attractive. Instead, they told us that the NGNP prototype reactor should be located where the petrochemical or other industries that use process heat could benefit from it. This is consistent with a finding in the Nuclear Energy Advisory Committee's 2011 Phase 1 review of the NGNP Project. In its review, the committee stated that the business case to optimize NGNP use for process heat applications and electricity indicates that a site in proximity to a wide range of industrial uses is more appropriate and that a siting at the Idaho National Laboratory will not support a partnership agreement with industry. If industry cannot realize an economic benefit from the prototype reactor, it is unlikely that industry would support the reactor being built at the Idaho National Laboratory.

- **Fiscal constraints and competing priorities.** NE officials, laboratory staff, industry representatives, and Nuclear Energy Advisory Committee members that we interviewed told us that NE's recent funding levels are inadequate to move forward with the NGNP prototype reactor. NE officials and the NGNP Industry Alliance both estimate that NE's share of NGNP Project could amount to as much as $2 billion over the remainder of the project, in which costs would be shared with industry. Under the NGNP Industry Alliance's proposal, DOE would provide between approximately $170 million and $330 million annually over the first 6 years of the proposed plan. This compares to the total funding for Advanced Reactor Concepts subprogram of about $60 million in 2014. NE officials, Nuclear Energy Advisory Committee members, and laboratory staff told us that NE's funding levels are inadequate to move forward toward a prototype reactor, even if it were to focus its resources on one effort. Furthermore, an NE officials and a member of the Nuclear Energy Advisory Committee told us that current priorities to fund R&D aimed at sustaining the existing light water reactors and focusing on the design and licensing of small modular reactors would have to shift in order to make more funds available for advanced reactor R&D.
- **Competition from natural gas.** NE officials, some industry representatives, and Nuclear Energy Advisory Committee members that we interviewed told us that low natural gas prices have made nuclear energy less attractive economically over the past few years, reducing overall interest in nuclear power options. NE officials and industry representatives said that the current atmosphere is not conducive to partnering with industry on advanced reactor projects, including the NGNP Project.

The Secretary's October 2011 letter to Congress did not specify which conditions might warrant a change in program direction—that is, proceed with Phase 2 of the NGNP Project—and NE has not developed a strategy for overcoming the identified barriers hindering the resumption of the project or a set of criteria for determining when a change in program direction would occur. An NE management official that we interviewed stated that conditions that would warrant a change in direction might include Congress legislating a carbon tax, a rise in price of natural gas, or an increase in funding for the NGNP. In addition, developing such a strategy may involve consultation with the Nuclear Energy Advisory Committee and others, including independent nuclear experts. Without a strategy for overcoming the barriers hindering the restart of the NGNP Project and identifying conditions NE can use for determining when a change in program direction would occur, it will be difficult for NE to demonstrate that it is poised to move forward, and it risks the project being on hold indefinitely.

According to EPAct 2005, NE was required to select the initial reactor design parameters to be used for the NGNP Project by September 30, 2011, or submit a report to Congress establishing an alternative date for making the selection. However, the Secretary's 2011 letter to Congress did not specify initial design parameters for the NGNP or specify an alternative date for making a selection. Instead, the letter stated that the initial design parameters had not yet been selected and that such a selection would be made by the public-private partnership once it is formed. Without selecting initial reactor design parameters or establishing a date to make a selection as required by EPAct 2005, it is not clear if or when NE is going to take this next step in deploying the NGNP prototype reactor.

In addition, after the Secretary's decision not to proceed with Phase 2 design activities, NE's engagement with NRC on licensing issues has decreased. NE maintains a team seeking to engage NRC on NGNP licensing issues, according to NE officials, but NRC has reassigned staff from its NGNP work and engages with NE on a minimal basis, according to NRC officials. NRC officials said that they cannot proceed substantively further in developing a licensing framework until NE has developed a specific design for an advanced reactor technology.

Furthermore, not deploying an advanced reactor prototype carries some risks, according to some industry representatives, Nuclear Energy Advisory Committee members, and NE officials we interviewed. Specifically, these risks include (1) falling behind other countries in advanced reactor development and losing market share in the global market for nuclear energy; (2) losing influence on which reactor technologies are developed, which raises safety and nuclear proliferation concerns; and (3) losing its ability to manufacture the components necessary to construct nuclear plants.

By not deploying an advanced reactor prototype, the United States risks falling behind other countries—such as Japan, Russia, China, South Korea, and France—that are actively working to deploy and commercialize advanced reactors, according to a Nuclear Energy Advisory Committee report. For example, Russia currently has two sodium-cooled fast reactors—one experimental and one commercial—in operation and is developing or constructing additional sodium-cooled fast reactor technologies, and it has plans to export reactor technology to other nations. In addition, China has an operating sodium-cooled fast reactor and high-temperature gas-cooled reactor on a test reactor scale and is in the process of building a prototype high temperature gas-cooled reactor, according to NE officials. In potentially losing its global leadership position in developing nuclear technologies, the United States risks losing market share in the global market for nuclear energy, which would cost the U.S. economy high-paying jobs in the nuclear industry, according to these individuals.

By falling behind other countries in advanced reactor development, the United States also risks losing influence on determinations of which reactor technologies are developed, with implications for the safety of reactor operations worldwide, as well as implications for how resistant the technologies are to nuclear proliferation—including the safe, effective disposal of nuclear waste—according to NE officials, laboratory staff, and Nuclear Energy Advisory Committee members that we spoke to. Specifically, according to members of the Nuclear Energy Advisory Committee, if the nation is not leading the development of advanced reactors, other countries may operate reactors that do not meet the highest safety standards and may not take adequate steps to ensure nuclear waste is handled appropriately and properly secured.

Similarly, the United States risks losing its ability to manufacture the components necessary to construct nuclear plants, according laboratory staff and Nuclear Energy Advisory Committee members that we interviewed. In addition, by not deploying an advanced reactor, NE risks losing staff, including engineers with the knowledge and experience necessary to design and build advanced reactors, according to an NE official and laboratory staff. For example, without a specific goal of developing an advanced reactor prototype, NE staff are more likely to leave NE for jobs with a better sense of mission, according to these officials.

CONCLUSIONS

Energy demand in the United States is expected to rise considerably over the coming decades, and concerns remain over energy security and greenhouse gas emissions from the burning of fossil fuels. While nuclear energy accounts for about 20 percent of electricity generation in the United States and produces no air pollution or greenhouse gases, the accident at Japan's Fukushima Daiichi commercial nuclear power plant in March 2011 highlighted ongoing concerns about the safety of nuclear plants, and concerns also exist about potential threats of nuclear proliferation and terrorism. With this in mind, it is important that nuclear power plants continue to evolve and provide energy economically, while also addressing safety and proliferation concerns. By conducting nuclear reactor R&D, NE has a critical role to play as it supports existing light water reactors, as well as a new generation of advanced nuclear reactors.

However, in 2011, DOE informed Congress that it would not proceed with Phase 2 of the NGNP Project until circumstances warranted a change in direction, and the project remains hindered by several barriers, including the cost-share and site requirements of EPAct 2005. NE officials have attempted to work out a cost-share agreement with the alliance, but different interpretations of the cost-share requirements by DOE and the NGNP Industry Alliance have created an impasse, and no agreement had been reached before DOE determined that it would not proceed to Phase 2 of the project.

Another barrier to proceeding with the project is the EPAct 2005 requirement that the NGNP reactor prototype be located at the Idaho National Laboratory. Building the reactor there foregoes the economic benefit of the reactor's process heat because industries that could potentially use the prototype reactor's high-temperature process heat are not located near the laboratory. If industry cannot realize an economic benefit from the prototype reactor, it is unlikely that industry would support the reactor being built at the Idaho National Laboratory.

Moreover, DOE's October 2011 letter notified Congress that the department had selected the NGNP technology, as required by EPAct 2005, acknowledged that the department had not selected the initial design parameters for the NGNP or identified the date upon which it would do so by September 30, 2011, as required by EPAct 2005, and essentially put Phase 2 of the NGNP Project on hold until conditions favorable to completing the NGNP warranted a change in direction. However, the letter did not specify which conditions might warrant a change in program direction—that is, proceed with Phase 2 of the NGNP Project—and NE has not developed a strategy for overcoming the identified barriers hindering restarting the project or which contains such conditions. Without a strategy for overcoming the barriers hindering the restart of the NGNP Project and identifying conditions NE can use for determining when a change in program direction would occur, the project may be on hold indefinitely. Furthermore, without selecting initial reactor design parameters and reporting the parameters to Congress, as required by EPAct 2005 for completing Phase 1 of the project, or establishing a date to make a selection, it is not clear if or when NE is going to take this next step and proceed with Phase 2 of the NGNP Project.

RECOMMENDATIONS FOR EXECUTIVE ACTION

To better prepare the Department of Energy to meet the requirement of the Energy Policy Act of 2005 to deploy the NGNP prototype reactor, we recommend that DOE take the following two actions:

- Develop, in consultation with the Nuclear Energy Advisory Committee and independent nuclear experts, as appropriate, a strategy to proceed with Phase 2 of the NGNP Project, outlining conditions that will warrant a change in program direction, remaining research and development activities, projected project budget and schedule, and steps necessary to overcome barriers to successful completion of the NGNP Project.
- Provide a report to Congress complying with the statutory requirement for Phase 1 of the NGNP Project by providing initial design parameters or a date for providing them. The report should also provide an updated status of the issues DOE identified in its 2011 letter to Congress and outline any additional barriers to proceeding with Phase 2 of the project, including
  - the status of the establishment of a public-private partnership;
  - the project strategy, including conditions that would warrant restarting the project; and
  - a legislative proposal, if necessary, to address any barriers to proceeding with the project, including the site and cost-share requirements.

AGENCY COMMENTS AND OUR EVALUATION

We provided a draft of this report to DOE for review and comment. In written comments, DOE's Assistant Secretary for Nuclear Energy, responding on behalf of DOE, wrote that DOE agreed in principle with our first recommendation and respectfully disagreed with our second recommendation. In addition, DOE provided technical comments, which we incorporated in the report as appropriate.

In its comment letter, DOE stated that it agreed in principle with our recommendation that it develop a strategy to proceed with Phase 2 of the NGNP Project. Moreover, DOE stated that its current strategy is to continue updating analyses of requirements for successful commercialization of reactor technology to reflect changing market conditions, research and development accomplishments, and the maturity of the licensing framework. However, this strategy does not outline steps DOE could take to proactively overcome the barriers hindering the resumption of the NGNP Project, nor does it outline criteria for determining when a change in program direction would occur. We continue to believe that developing a strategy to proceed with Phase 2 of the NGNP Project is important because without a strategy it will be difficult for NE to demonstrate that, upon completion of Phase 1, it will be poised to develop a final design and construct and operate the prototype reactor. Moreover, not having a strategy for proceeding with Phase 2 could result in the project being on hold indefinitely.

DOE respectfully disagreed with our recommendation that it provide a report to Congress that, among other things, provides initial design parameters or a date for providing them and outlines barriers to proceeding with Phase 2 of the project. DOE stated that such a report was not advisable or useful, or necessary as a means for the Department to comply with the statutory requirements for Phase 1 of the NGNP Project, and further stated that the Department is in compliance with the relevant statutory requirements. As DOE explained, it reported to Congress in 2011 that while it had selected the hydrogen production technology, as required by EPAct 2005, it had not selected the initial design parameters for the project and that based on the recommendations of the Nuclear Energy Advisory Committee, fiscal constraints, competing priorities, projected cost of the prototype, and the inability to reach agreement with industry on cost sharing, it would not proceed with Phase 2 design activities at that time. Instead, it would continue to focus on high-temperature reactor R&D activities and establishment of a public-private partnership, among other things, until conditions warranted a change in direction. DOE did not, however, establish an alternative date for selecting the initial design parameters, as EPAct 2005 required. Rather, it stated that selection of initial design parameters would be made by the public-private partnership once it is formed. Given that almost 3 years have passed since the letter to Congress, we believe that the recommended report is warranted and would serve to inform Congress of the status of the NGNP Project and provide transparency and accountability regarding DOE's intentions for completing Phase 1 and proceeding with Phase 2 of the project. For example, by providing a firm date for selecting the initial design parameters of the NGNP prototype reactor, DOE could be held accountable to meeting that date or could engage in a discussion about whether and why that date should be further extended. Similarly, an updated report to Congress could include a candid description of the ongoing barriers to moving forward, which could spur discussions resulting in legislation or other remedies to mitigate these barriers.

Sincerely yours,

Frank Rusco
Director, Natural Resources and Environment

Timothy M. Persons, Ph.D., Chief Scientist
Director, Center for Science, Technology, and Engineering

Appendix I: Overview of Nuclear Reactor Technologies

Overview of Reactor Operations

Nuclear reactors generate heat by sustaining a fission chain reaction in nuclear fuel. Nuclear fission reactions can occur when a neutron strikes the nucleus of a large atom, causing that nucleus to split, or fission. The result of a fission reaction is typically two fission fragments, or smaller nuclei; 2 or 3 new fast-moving neutrons; and significant heat. In a nuclear reactor, the large atoms used for fission are typically the fissile isotopes uranium-235 or plutonium-239, and the new neutrons produced by a fission reaction are used to initiate new fission reactions, resulting in a sustained fission chain reaction.[36]

This heat generated by this fission reaction is typically used to create steam and drive a steam turbine to generate electricity. Some reactors may also operate at particularly high temperatures and can use the heat to either generate electricity or to supply process heat that can be used for various industrial processes, replacing other heat sources such as natural gas.

Types of Nuclear Reactors

Nuclear reactors typically fall into one of two types based on the neutron spectrum, or neutron energies, at which the fission reactions occur as follows:

- Thermal reactors optimize the fission reaction rate in their fuel. This is done by slowing down, or moderating, the high-energy fast neutrons that are the products of fission reactions. This thermalization of the fast neutrons increases the likelihood that a neutron will initiate a fission reaction. Currently deployed light water reactors, including pressurized water reactors and boiling water reactors, are thermal reactors.
- Fast reactors, by contrast, do not moderate the fission neutrons, leaving them as fast neutrons. A fast neutron has a lower likelihood of initiating a fission event than a slow neutron, so the chain reaction can be more difficult to sustain, but it has the benefit of producing more neutrons when fission does occur. These surplus neutrons, as compared to the number of neutrons produced in thermal reactors, allow fast reactors to be more effective than thermal reactors at creating, or breeding, new fuel through neutron bombardment of uranium-238 (creating plutonium-239). Fast reactors optimized for fuel production in this manner are called fast breeder reactors and can produce more fuel through breeding than they consume. Fast reactors may also use spent fuel from other nuclear reactors as fuel and thereby reduce long-term fuel disposal needs.

Overview of the Reactor Core

While there are a large number of reactor technologies that can differ significantly, the fission reaction in a reactor occurs in the central region of a reactor called the reactor core. The reactor core typically contains several components as follows:

- **Nuclear fuel.** Nuclear reactors need fissile isotopes, such as uranium-235 and plutonium-239, to sustain chain reactions. Commercial reactors often use uranium ore that has been enriched in the isotope uranium-235 as their fissile fuel; the rest of the fuel consists of the non-fissile uranium-238. However, reactor operation will result in the conversion of some uranium-238 to the fissile isotope plutonium-239, which may then fission and contribute to power generation, and some reactor fuel may start with some of the uranium-235 mixed with plutonium-239 (sometimes referred to as a mixed oxide fuel). Fast reactors can also use spent fuel from other reactors as fuel and can be very effective at converting uranium-238 into plutonium-239. Some reactors can also utilize uranium-233 or thorium-232 as components of their fuel.
- **Moderator.** Thermal reactors use a moderator material to slow down, or thermalize, the fission neutrons in order to sustain the fission reaction. This is needed because neutrons produced by fission reactions are too fast (or energetic) to have a high likelihood of initiating a new fission reaction in fuel. Fast reactors are designed to utilize fast neutrons for the fission reactions and fuel breeding and, as such, do not use a moderator.
- **Coolant.** To remove heat from the core, a coolant—typically water, a gas, or liquid metal—is circulated through the core. The coolant both prevents the core from overheating (which could damage or melt the fuel) and it carries energy, in the form of heat, outside the core for electricity production, typically by generating steam that then drives a steam turbine. In some reactor types, the coolant can also function as the reactor's moderator.
- **Reaction control.** Reactors can use different techniques to maintain the fission chain reaction at appropriate rates. For example, control rods may be inserted into reactor cores to absorb neutrons and slow down (or stop) the chain reaction, or neutron-absorbing materials, such as boric acid in pressurized water reactors, may be introduced to the coolant system to achieve a similar effect.

Table 2. Characteristics of Commercial Reactor Types, including Those Currently Operating in the United States and Advanced Reactor Designs Under Development

Reactor type (italics indicate current commercial deployment in the United States)	Fuel	Coolant	Moderator	Process heat	Neutron spectrum
Light water reactors • *Pressurized water reactors* • *Boiling water reactors*	Uranium	Water	Water	No	Thermal
Supercritical water cooled reactor	Uranium or mixed oxide	Supercritical water	Supercritical water	Yes	Thermal (may be designed as fast)
High-temperature gas-cooled reactor/very high temperature reactor	Uranium	Helium	Graphite	Yes	Thermal
Liquid-metal-cooled fast reactor • sodium-cooled fast reactor • lead-bismuth cooled fast reactor	Uranium, plutonium, or mixed oxide	Liquid metals (sodium, lead, etc.)	N/A	No	Fast
Gas-cooled fast reactor	Uranium, plutonium, or mixed oxide	Helium or carbon dioxide	N/A	No	Fast
Molten salt reactor or molten salt breeder reactor	Uranium, thorium, or mixed oxide	Molten salts	Graphite (none for breeder)	Yes	Thermal (may be designed as fast)

Sources: GAO analysis of DOE, International Atomic Energy Agency (IAEA), and other documents. | GAO-14-545

Reactor technologies are classified as either thermal or fast reactors (although some technologies are "epithermal" and fall in between the two types) and by the materials used for the moderator or coolant. For example, a pressurized water reactor is a thermal reactor using water as both a coolant and moderator, and a gas-cooled fast reactor is a fast reactor using gas (carbon dioxide or helium) as a coolant.

Table 2 lists and provides information about the reactor types that are either currently operating in the United States or are advanced reactor designs under consideration for development.

End Notes

[1] Spent (or used) nuclear fuel—fuel that has been used in a reactor and removed from the reactor core—from light water reactors in the United States is currently stored at nuclear reactor sites around the country. About three-quarters of the fuel is stored in pools of water, and about one-quarter is stored in dry storage casks. For more information, see GAO, *Spent Nuclear Fuel: Accumulating Quantities at Commercial Reactors Present Storage and Other Challenges*, GAO-12-797 (Washington, D.C.: Aug. 15, 2012).

[2] Consistent with the Treaty on Nonproliferation of Nuclear Weapons, nuclear proliferation includes the transfer of nuclear weapons, related technology, or special fissionable material to nonnuclear weapons states. One factor contributing to the risks of nuclear proliferation and terrorism is the challenge of protecting spent nuclear fuel from diversion and theft.

[3] Light water reactors, also called Generation-II and Generation-III reactors, use ordinary (i.e., "light") water, either pressurized or boiling, for reactor moderation and cooling of the reactor core. Advanced reactors, also called Generation-IV reactors, use substances such as liquid metals or high-temperature gases to cool the reactor core. A moderator in a nuclear reactor is a substance that reduces the speed of the fast neutrons to better sustain the nuclear reaction. For more technical details of reactor types, see appendix I.

[4] For these reactors to continue operating, their owners must renew their licenses with the Nuclear Regulatory Commission (NRC), the federal agency responsible for licensing and regulating civilian nuclear reactors. In general, NRC licenses reactors to operate for 40 years and allows reactor owners to apply to renew their operating licenses for up to an additional 20 years per renewal. As of April 2014, 72 of the 100 currently operating commercial nuclear power reactors in the United States are operating with renewed operating licenses.

[5] The U.S. administration's goal is to reduce greenhouse gas emission levels by 83 percent of 2005 levels by the year 2050.

[6] GAO, Nuclear Energy: Status of DOE's Effort to Develop the Next Generation Nuclear Plant, GAO-06-1056 (Washington, D.C.: Sept. 20, 2006).

[7] The Nuclear Energy Advisory Committee includes representatives from universities, industry, foreign nationals, and national laboratories. The committee paid particular attention to obtaining a diverse membership with a balance of disciplines, interests, experiences, points of view, and geography. The committee operates in accordance with the Federal Advisory Committee Act (FACA) and all applicable FACA amendments, federal regulations, and Executive Orders.

[8] See, for example, Nuclear Energy Advisory Committee, *Report to Nuclear Energy Advisory Committee Fuel Cycle Subcommittee Meeting of April 23, 2013* (Washington, D.C.: June 13, 2013); GAO, *Nuclear Fuel Cycle Options: DOE Needs to Enhance Planning for*

Technology Assessment and Collaboration with Industry and Other Countries, GAO-12-70 (Washington, D.C.: Oct. 17, 2011); GAO, *Global Nuclear Energy Partnership: DOE Should Reassess Its Approach to Designing and Building Spent Nuclear Fuel Recycling Facilities*, GAO-08-483 (Washington, D.C.: Apr. 22, 2008); *National Academies Press, Review of DOE's Nuclear Energy Research and Development Program* (Washington, D.C.: 2008).

[9] Because this was a nonprobability sample, the information gathered from these companies cannot be generalized to the nuclear power industry as a whole, but it can provide illustrative examples. Based on our review of companies developing advanced reactor technology, we selected these companies because they had involvement with DOE's advanced reactor R&D efforts.

[10] The Nuclear Energy Advisory Committee provides independent advice to the Office of Nuclear Energy on complex science and technical issues that arise in the planning, managing, and implementation of DOE's nuclear energy program. The National Research Council's mission is to improve government decision making and public policy, increase public understanding, and promote the acquisition and dissemination of knowledge in matters involving science, engineering, technology, and health. The Next Generation Nuclear Plant Industry Alliance is an alliance of nuclear technology companies, electric utilities, and industrial end users that seek to support efforts to design, build, operate, and use high-temperature gas-cooled reactors. The Nuclear Energy Institute develops policy on legislative and regulatory issues affecting the nuclear industry and serves as the nuclear industry's voice before the U.S. Congress, executive branch agencies, and federal regulators.

[11] The DOE Office of Nuclear Energy provides funding to 10 national laboratories: (1) Argonne, (2) Brookhaven, (3) Idaho, (4) Lawrence Berkeley, (5) Lawrence Livermore, (6) Los Alamos, (7) Oak Ridge, (8) Pacific Northwest, (9) Sandia, and (10) Savannah River.

[12] The Advanced Reactor Concepts subprogram will be renamed the Advanced Reactor Technologies subprogram beginning in fiscal year 2015.

[13] Beginning in fiscal year 2015, the Advanced Small Modular Reactor subprogram will be consolidated into the Advanced Reactor Technologies subprogram.

[14] Sustainable fuel cycle options are those that would better utilize uranium resources, maximize energy generation, minimize waste generation, improve safety, and limit proliferation and terrorism risks, according to DOE.

[15] Accident-tolerant fuels are fuels that offer improved reliability and safety characteristics during accident conditions. Proliferation-resistant fuels are those that would make it more difficult or time-consuming to weaponize.

[16] In addition to these programs, NE funds programs that coordinate its international activities, manage and protect its facilities, and provide overall direction and execution of NE's programs.

[17] NE conducts most of the advanced nuclear R&D at DOE's national laboratories, but some NE-funded R&D is carried out at university laboratories, and some is done by private industry through cooperative agreements.

[18] NE's Nuclear Energy University Program was established in 2009 to consolidate university support under one initiative and better integrate university research within NE technical programs. The mission of the Nuclear Energy University Program is to engage U.S. universities in conducting R&D that supports DOE's mission, improving related infrastructure, and supporting student education.

[19] As of January 2014, the 13 participating members were Argentina, Brazil, Canada, China, Euratom (the implementing organization for development of nuclear energy within the European Union), France, Japan, Russia, South Korea, South Africa, Switzerland, the United Kingdom, and the United States, according to the Organisation for Economic Co-operation and Development (OECD).

[20] The high-temperature gas-cooled reactor is an underground helium-cooled nuclear reactor technology. The reactor and the nuclear heat supply system consist of three major components: the reactor, a heat transport system, and a cross vessel that routes the helium between the reactor and the heat transport system.

[21] Advanced reactors have been built and tested in countries around the world, including in the United States, since the 1950s. Sodium-cooled fast reactors have been built and operated for research and commercial purposes in countries around the world, including France, Japan, Germany, Russia, the United Kingdom, and the United States. Two notable U.S. fast reactors include the Experimental Breeder Reactor, operational from 1964 through 1994 and the Fast Flux Test Facility, operational from 1982 to 1992. High-temperature gas-cooled reactors have also been built and operated around the world, including two that were licensed and operated in the United States. One reactor—Peach Bottom Atomic Power Station, Unit 1—was located in Pennsylvania and operated from 1967 to 1974, according to the U.S. Nuclear Regulatory Commission. The other—Fort St. Vrain Generating Station—was located in Colorado and operated between 1979 and 1989.

[22] Pub. L. No. 109-58, §§ 641-45, 119 Stat. 594, 794-99 (2005) (codified at 42 U.S.C. §§ 16021-16025).

[23] Nuclear Energy Research Advisory Committee, *A Review of the NGNP Project* (Feb. 22, 2006).

[24] Engineering-scale demonstrations typically precede commercial-scale deployment and are meant to ensure technologies work as intended before an investment is made in a larger plant.

[25] GAO-08-483.

[26] Under the accelerated approach, DOE would partner with industry and use the latest commercially available technology to design and build a commercial-scale reprocessing plant and advanced reactor without first building engineering-scale facilities.

[27] National Research Council, *Review of DOE's Nuclear Energy Research and Development Program* (2008).

[28] EPAct 2005 also provides that DOE may submit a report to Congress establishing an alternative date for completion.

[29] The March 2012 funding opportunity announcement had a target date of 2022 for deployment. The March 2013 funding opportunity announcement had a target date of 2025 for deployment.

[30] Spent nuclear fuel can no longer efficiently generate power in a nuclear reactor, but it can be reprocessed to separate out uranium and plutonium which may be used again to fuel a reactor. Reprocessing, however, results in nuclear waste that requires disposal. The United States does not currently reprocess its spent nuclear fuel, and this fuel, when it is accepted for disposal, is considered to be high-level waste as defined by NRC.

[31] The Nuclear Waste Policy Act of 1982 directed DOE to investigate sites for a federal geologic repository to dispose of spent nuclear fuel and high-level nuclear waste from commercial nuclear power plants. In 1986, DOE recommended three candidate sites for characterization, but in 1987, the Nuclear Waste Policy Amendments Act of 1987 directed DOE to focus its efforts only on Yucca Mountain, a site about 100 miles northwest of Las Vegas, Nevada. In March 2009, the Secretary of Energy announced plans to terminate the Yucca Mountain repository program. See GAO-12-797 for more information. The administration's current strategy proposes to move spent fuel from reactor sites to an interim storage facility for eventual disposal in a geological repository underground.

[32] The International Atomic Energy Agency is an autonomous international organization affiliated with the United Nations, established in Vienna, Austria, in 1957. The agency has the dual role of promoting the peaceful uses of nuclear energy by transferring nuclear science and technology through its nuclear science and applications and technical cooperation programs, and verifying, through its safeguards program, that nuclear material subject to safeguards is not diverted to nuclear weapons or other proscribed purposes.

[33] The OECD is an organization of 34 industrialized countries, operating by consensus, that fosters dialogue among members to discuss, develop, and refine economic and social policies and provides an arena for establishing multilateral agreements.

[34] Current reprocessing techniques, aimed at extracting usable uranium and plutonium from spent fuel, have been developed for use in light water reactors. However, spent fuel from light water reactors contains other radioactive isotopes that a fast reactor can eliminate from the waste stream.

[35] As discussed above, some advanced reactor technologies have more significant technical and economic barriers to overcome, and thus are not realistic candidates for deployment in the near term. For instance, while NE could prioritize the sodium-cooled fast reactor, some industry representatives and members of the National Academy of Sciences' National Research Council told us that a fast reactor technology market does not exist in the United States. Other advanced reactors, such as fluoride salt-cooled high-temperature reactors and lead-cooled fast reactors, are not very mature or well tested, or have unproven technologies and components, according to NE officials.

[36] Isotopes are varieties of a given chemical element with the same number of protons but different numbers of neutrons. For example, the helium-3 isotope, which is used in research and to detect neutrons in radiation detection equipment, has one less neutron than the helium-4 isotope, which is the helium isotope commonly used in party balloons.

Index

D

E

V

W

Z